Diversity and Ecology of Invasive Plants

Edited by Sudam Charan Sahu and Sanjeet Kumar

Published in London, United Kingdom

IntechOpen

Supporting open minds since 2005

Diversity and Ecology of Invasive Plants
http://dx.doi.org/10.5772/intechopen.73424
Edited by Sudam Charan Sahu and Sanjeet Kumar

Contributors
Mia Maltz, Brooke Pickett, Emma Aronson, Wanessa Almeida Da Costa, Sérgio Henrique Brabo De Sousa, Priscila do Nascimento Bezerra, Mozaniel Santana de Oliveira, Fernanda Wariss Figueiredo Bezerra, Jorddy Neves Da Cruz, Sebastião Gomes Silva, Renato Macedo Cordeiro, Cintya Cordovil Rodrigues, Antônio Robson Batista De Carvalho, Daniel Santiago Pereira, Antonio Pedro Da Silva Souza Filho, Raul Nunes de Carvalho Jr, Cinthya Elen Pereira de Lima, Pedro Alam de Araújo Sarges, Pablo Becerra, Ramiro Bustamante, Anneliese Fuchs, Christina Pichler-Koban, Michael Jungmeier, Wilfried Elmenreich, Sudam Charan Sahu

First published in London, United Kingdom, 2019 by IntechOpen
IntechOpen is the global imprint of INTECHOPEN LIMITED, registered in England and Wales, registration number: 11086078, 7th floor, 10 Lower Thames Street, London, EC3R 6AF, United Kingdom
Printed in Croatia

British Library Cataloguing-in-Publication Data
A catalogue record for this book is available from the British Library

Additional hard and PDF copies can be obtained from orders@intechopen.com

Diversity and Ecology of Invasive Plants
Edited by Sudam Charan Sahu and Sanjeet Kumar
p. cm.
Print ISBN 978-1-83968-351-0
Online ISBN 978-1-83968-352-7
eBook (PDF) ISBN 978-1-83968-353-4

For EU product safety concerns:
IN TECH d.o.o., Prolaz Marije Krucifikse Kozulić 3, 51000 Rijeka, Croatia, info@intechopen.com or visit our website at intechopen.com.

Meet the editors

Sudam Charan Sahu, M.Sc., PhD, F.I.A.T. is currently working as an Assistant Professor at the Department of Botany, North Orissa University, Baripada (Odisha), India. He completed his PhD from CSIR-IMMT, Bhubaneswar (Under Utkal University, Bhubaneswar), India and his Post-Doctorate from Indian Institute of Science, Bangalore (India). His specialization and research fields include plant taxonomy, ethnobotany, forest ecology, climate change, and biodiversity conservation. He has published more than 40 papers in various national and international journals, one book, 12 book chapters and edited 2 books. He is recognized as a Fellow of the Indian Association for Angiosperm Taxonomy. He was also awarded with the DST-Young Scientist from Science & Engineering Research Board (SERB), DST, Government of India. He is a well-recognized reviewer of many SCI and non-SCI journals.

Dr. Sanjeet Kumar, PhD, DELF, is a founder and CEO of Ambika Prasad Research Foundation, Odisha. His current research is focused on status, taxonomy, diversity, phytochemistry; wetland ecosystems and their biowealth; establishment of relationships between flora and fauna, and restoration of medicinal plants found in riverine ecology. He has 10 years of research experience on medicinal plants. He has discovered 4 new orchid species for the flora of Manipur and 3 species new to India. He has about 100 publications in journals of national and international repute. He has written 8 books and is currently writing 2 books on medicinal plants. He has organized two national seminars on medicinal plants in association with the Ministry of AYUSH, Government of India.

Contents

Preface

The study on the diversity and ecology of invasive plants in the recent Anthropocene era has enormously increased. Invasive species are one of the worst threats to biodiversity and ecosystems across the globe. According to the Rio Convention of Biological Diversity (1992), invasive alien species were ranked the second highest threat to the loss of biodiversity. Invasive plants are non-native plants that are introduced into an area and spread at a very fast rate with significant ecological and economic impacts. In a new area, an invasive plant competes with native plants for light, nutrients, water, and other resources, thereby replacing the native vegetation. As a consequence, the biodiversity and natural beauty of the environment are distorted. Controlling invasive plant species is very difficult and costly. There are various methods available for control such as mechanical, chemical, and biological. Further, climate change exacerbates the spread of invasive alien plant species. Therefore, it is important how to control it by applying modern tools and techniques.

This book gives an account of diversity, distribution, and ecology of invasive plant species around the globe, their impact on ecosystem, and control and management for conserving native biodiversity. Section I (Diversity of invasive species) describes the diversity and distribution of invasive plant species in various parts of the world. Section II (Impact of invasive species on ecosystem and its control) sheds light on the impact of invasive plants on natural and artificial ecosystems, economics and human health and also control and management for biodiversity conservation.

We express our deep gratitude to the authors for sharing their knowledge by contributing chapters to this book. We are extremely thankful to Ms. Dolores Kuzelj, Ms. Sara Petanjek, and other officials for their constant support during various phases of publication of this book and thank them for providing us opportunity to edit the book.

We hope that this book will be helpful to a wider group of people; researchers, scientists, forest managers, conservation biologists, and academicians.

Dr. Sudam Charan Sahu
Assistant Professor,
Department of Botany,
North Orissa University,
Baripada (Odisha),
India

Dr. Sanjeet Kumar
Ambika Prasad Research Foundation,
Bhubaneswar (Odisha),
India

Section 1

Diversity of Invasive Species

Chapter 1

Invasive Alien Flora in and around an Urban Area of India

Samarendra Narayan Mallick, Nirius Xenan Ekka, Sanjeet Kumar and Sudam C. Sahu

Abstract

Invasive alien species are non-native exotic organisms which can disperse and destroy the biodiversity and change the ecosystem. The present study deals with the comprehensive list of invasive alien plants (IAPs) of Rourkela Steel City, Sundargarh, Odisha, with background information on family, habit, and nativity. A total of 165 invasive alien species under 132 genera and 59 families have been recorded. From the nativity study, among 25 geographic regions, the majority of invasive plants reported from American continent (62%) with 103 species. While in life form analysis, the herbs (114 species) are dominant, followed by trees (23 species), shrubs (22 species), climber (5 species), and undershrub (1 species). *Ageratum conyzoides*, *Blumea lacera*, *Cassia alata*, *Lantana camara*, *Cassia tora*, *Parthenium hysterophorus*, *Xanthium* sp., *Datura* sp., *Cardamine scutata*, *Argemone mexicana*, *Grangea maderaspatana*, *Hyptis suaveolens*, and *Gnaphalium polycaulon* are some noxious species found during the study. *Parthenium hysterophorus* is the highly noxious plant which is grown everywhere after *Ageratum conyzoides* and *Lantana camara*. Most of the invasive species are locally used for medicinal purposes as well as for food, fuel, and fodder purposes. A better planning and reporting of the spread of new plants in the area are needed for early identification and control of the invasive alien plant species in different seasons. Since the flora of Sundargarh districts has not been beneficially explored, this study will help in the compilation of flora of Sundargarh district and Rourkela in particular. Further studies will reveal the allelopathic effects on different agricultural crops as well as the different ethnobotanical values.

Keywords: invasive alien plants, biodiversity, utility, urban area, India

1. Introduction

Human beings depend on plants for his daily needs for which several numbers of plants are used to fulfill their purposes. Sometimes to fulfill human needs, plants are introduced intentionally by humans or accidentally from one region to another new region which is nonnative. These introduced plant species are called alien species or exotic species. The alien species invade the new region after well adapted to the environment. The plants which are introduced by human intentionally or accidentally by migration from its natural habitat to another new habit and their localities are known as alien, introduced, and exotic, originated from foreign or nonnative species [25, 33]. They have the potential to grow in any environmental conditions and are easily invasive to the new environment. Preston and Williams

[22] stated that "Invasive alien plant species (IAPs)" are grown in such a way that they become as more dangerous to sustainable development. As a result, we are facing the great challenge of biodiversity loss all over the globe. These group of plant species act as the main cause for threat to the native biological diversity. They show various effects on the environment and economy of nonnative ecosystems. The exotic or alien plant species not only show negative impacts, but also they have much economic benefits. Now invasion alien species are cultivated to provide food, medicine, fuel, or fodder to local communities [9, 29]. The international trade of the products is helpful for introduction of these invasive alien species. Globally the introduction of IAPs leads to the huge loss of biodiversity and agriculture crops and health problems like respiratory illness [19].

Invasion of plants creates serious problems to the ecosystems by changing the structure, composition, and function of natural ecosystem [15, 17]. The rapid reproduction and growth rate, high dispersal ability, physiological adaptations to new conditions, and ability to survive on various ecosystems are the common characteristics of invasive plants. The IAPs have the ability to associate with human beings very easily. When the invasive plants colonized to grow in new areas, it can change the soil structure and composition of that area. It is reported that the agricultural lands are more threatened by IAPs because they are introduced by the crop seeds, garden plants, and wind breakers [24].

The first and most important step for effective and proper management of IAPs is to collect the baseline data about their invasion status, growth form, and life cycle. Accurately distinguishing between native and alien species is required not only when developing conservation and vegetation management plans but also for improving our understanding of the different components of biodiversity [21]. Rourkela, one of the major steel industrial centers of India and regarded as the industrial capital of Odisha, is situated in the north-eastern part of the state. Rourkela is located in Sundargarh district about 245 km from the shoreline of Bay of Bengal. It is located at 20° 12′ North latitude and 84° 53′ longitude, at the elevation about 219 m above the mean sea level. Due to better communication, abundance of natural mineral resources such as iron ore, limestone, dolomite, water, and other infrastructures in and around Rourkela is the main reason for the starting of industrialization since 1956. Studies on flora of Sundargarh district have not been fully explored. A few reports on flora of Sundargarh district [1, 2, 11, 13, 14] have been published. The study of literatures reveals that survey pertaining to major invasive plant species has not been reported earlier. It is high time to undertake complete survey of the flora of Sundargarh district with special emphasis on IAPs which may not be available in the future due to rapid industrialization. Many species may become endangered in the process of development, and they should be recorded and identified along with their usefulness before their extinction during rapid industrialization. Keeping in view, an attempt has been made in the present study to provide the baseline information on the invasive plant species in and around of Rourkela City of Sundargarh district. It will be helpful in preparation of district flora of Sundargarh.

2. Materials and method

During January 2012 to April 2019, intensive floristic surveys were undertaken in different areas of Rourkela Steel City (**Figure 1**) in such a way that each location could be studied in every season of the year. A comprehensive list of invasive alien plant species (IAPs) and the interaction with local inhabitants were made to collect the information regarding the various uses of IAPs of the area. Periodic collection of IAPs was made from each locality followed by identification using the available

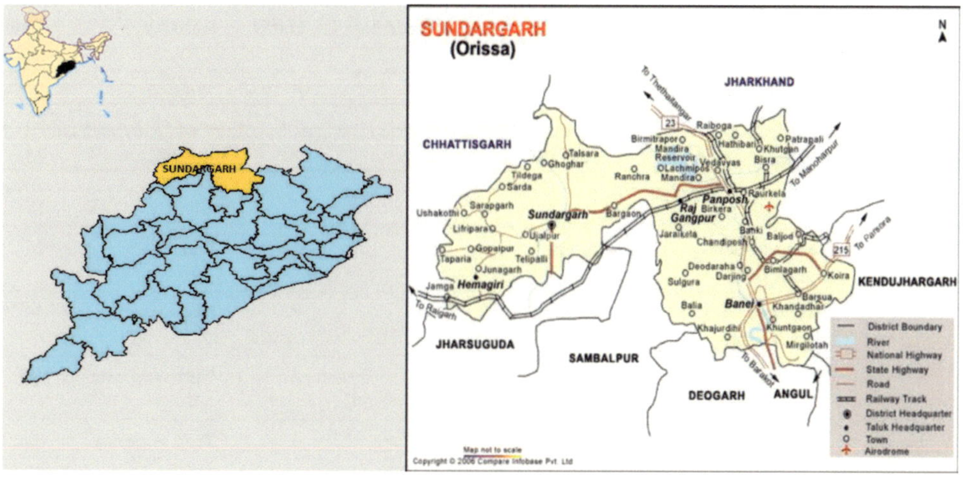

Figure 1.
Location map of Rourkela Steel City of Sundargarh district, Odisha, India.

floras [6, 30]. The nativity, history, diversity, sources, and mode of introduction of these alien invasive plants were noted from the available literatures. The native ranges of the species were recorded from published literatures [3, 4, 5, 10, 12, 15, 16, 18–20, 24–27, 31, 32, 36, 37, 38]. Plants were categorized according to their life forms as herb, undershrub, shrub, climber, and tree as well as their habit-wise as annual, biennial, and perennial. The studied habitats were wasteland, cultivated field, riverbank, pond bank, home garden, forest, roadside, etc. The economic importance of the IAPs was collected from the local inhabitants and surveyed literatures.

3. Results and discussion

A total of 165 taxa of invasive alien plant species belonging to 132 genera and 59 families have been recorded from the Rourkela Steel City of Odisha (**Table 1**). The number of dicotyledonous IAPs found is 149 under 118 genera and 50 families, while 15 species of monocotyledons are found under 14 genera and 8 families. From the study, it was found that 114 species (69%) were herbs followed by trees with 23 species (14%), shrub 22 species (13%), climbers 5 species (3%), and undershrubs 1 species (1%) (**Figure 2**). The life form pattern distribution showed that herbaceous species (114 spp.) were dominant than other life forms (**Table 2**). The herbs can easily grow in any condition of environment and dominate to others. The habit distribution analysis showed that 56% (92 spp.) were annuals and 44% (73 spp.) were perennials. **Table 3** showed the total number of IAPs recorded from the Rourkela Steel City and distributed under different families. From the taxonomic distribution of alien flora, Asteraceae (24 spp.) showed dominant impact among the invasive alien species in this region followed by Caesalpiniaceae (11 spp.), Convolvulaceae (9 spp.), Euphorbiaceae (8 spp.), Amaranthaceae (8 spp.), Poaceae (6 spp.), and Solanaceae (8 spp.), and Fabaceae, Malvaceae, and Verbenaceae represented only 5 spp. each (**Table 4**). These 10 dominant families contributed 89 species (54%) of the total invasive plant species studied (**Figure 3**). The genera *Cassia* and *Ipomoea* showed the highest number (six spp. each) followed by *Cleome*, *Euphorbia*, *Alternanthera*, *Ludwigia*, etc.

The contribution of different geographical regions or the nativity of invasive alien species is shown in **Table 3**. A total of 25 native geographical regions of IAPs were recorded. The major geographical regions or nativities of IAPs were Tropical

Sl no.	Plant species	Family	Life form	Habit	Nativity	Use
1	*Abelmoschus esculentus* Moench.	Malvaceae	Shrub	P	Trop. Africa	V, Ft
2	*Acacia auriculiformis* A. Cunn. ex Benth.	Mimosaceae	Tree	P	Australia	M, Ave, Sf
3	*Acanthospermum hispidum* DC.	Asteraceae	Herb	A	Brazil	M
4	*Aerva lanata* (L.) Juss. ex. Schult.	Amaranthaceae	Herb	P	Madagascar	M
5	*Aeschynomene indica* L.	Fabaceae	Herb	A	North America	Fu
6	*Ageratum conyzoides* L.	Asteraceae	Herb	A	Trop. America	Nox
7	*Allium cepa* L.	Liliaceae	Herb	A	Mediterranean	M, V
8	*Aloe barbadensis* Mill.	Liliaceae	Herb	P	Mediterranean	M
9	*Alternanthera philoxeroides* (Mart.) Griseb.	Amaranthaceae	Herb	P	Trop. America	V, M
10	*Alternanthera pungens* Kunth	Amaranthaceae	Herb	P	Trop. America	V
11	*Alternanthera sessilis* (Linn) DC.	Amaranthaceae	Herb	P	Trop. America	V, M
12	*Amaranthus spinosus* L.	Amaranthaceae	Herb	A	Trop. America	V
13	*Antigonon leptopus* Hook. & Arn.	Polygonaceae	Climber	P	Trop. S. America	O
14	*Argemone mexicana* L.	Papaveraceae	Herb	A	S. America (seventeenth cent.)	M, Nox
15	*Bauhinia purpurea* L.	Caesalpiniaceae	Tree	P	West Indies	V
16	*Bidens pilosa* L.	Asteraceae	Herb	A	Trop. America	M, Fo
17	*Blainvillea acmella* (L.) Philipson	Asteraceae	Herb	A	Trop. America (eighteenth)	M
18	*Blumea lacera* (Burm.f.) DC.	Asteraceae	Herb	A	Trop. America	Nox, M
19	*Borassus flabellifer* L.	Arecaceae	Tree	P	Trop. Africa	Ft, Fu
20	*Bougainvillea spectabilis* Willd.	Nyctaginaceae	Shrub	P	Brazil	O
21	*Caesalpinia pulcherrima* (L.) Sw.	Caesalpiniaceae	Shrub	P	Trop. America	O
22	*Calotropis gigantea* R. Br.	Asclepiadaceae	Shrub	P	Trop. Africa	M
23	*Calotropis procera* (Ait.) R. Br.	Asclepiadaceae	Shrub	P	Trop. Africa	M
24	*Cannabis sativa* L.	Cannabinaceae	Undershrub	P	Central Asia	M, Sm, Nar
25	*Capsicum annuum* L.	Solanaceae	Shrub	A	Trop. America	F

Sl no.	Plant species	Family	Life form	Habit	Nativity	Use
26	*Cardamine scutata* L.	Brassicaceae	Herb	A	Trop. America	NK, Nox
27	*Cardiospermum halicacabum* L.	Sapindaceae	Herb	A	Trop. America	M
28	*Carica papaya* L.	Caricaceae	Tree	P	Mexico	V
29	*Cassia alata* L.	Caesalpiniaceae	Shrub	P	West Indies	NK, Nox
30	*Cassia fistula* L.	Caesalpiniaceae	Tree	P	Pantropic	O, M, Sf
31	*Cassia obtusifolia* L.	Caesalpiniaceae	Herb	P	Trop. America	M
32	*Cassia occidentalis* L.	Caesalpiniaceae	Herb	P	S. America	M
33	*Cassia siamea* Lamk	Caesalpiniaceae	Tree	P	South East Trop. Asia	Fu, Ave
34	*Cassia tora* L.	Caesalpiniaceae	Herb	A	S. America (1824)	V, M, Nox
35	*Casuarina equisetifolia* Forster & Forster f.	Casuarinaceae	Tree	P	Australia	Fu, Sf
36	*Catharanthus pusillus* (Murr.) G. Don.	Apocynaceae	Herb	A	Trop. America	NK
37	*Catharanthus roseus* (Linn) G.Don	Apocynaceae	Shrub	P	West Indies	M
38	*Celosia argentea* L.	Amaranthaceae	Herb	A	Trop. Africa	V, M
39	*Chenopodium album* L.	Chenopodiaceae	Herb	A	Europe	V
40	*Chloris barbata* Sw.	Poaceae	Herb	P	Trop. America	Fo, Fu
41	*Chromolaena odorata* L.	Asteraceae	Herb	A	Trop. America	M
42	*Chrozophora rottleri* (Geisel.) Juss.	Euphorbiaceae	Herb	A	Trop. Africa	NK
43	*Cleome gynandra* L.	Capparaceae	Herb	A	Trop. America	M
44	*Cleome monophylla* L.	Capparaceae	Herb	A	Trop. Africa	V, M
45	*Cleome rutidosperma* DC	Capparaceae	Herb	A	Trop. America	NK
46	*Cleome viscosa* L.	Capparaceae	Herb	A	Trop. America	V, M
47	*Convolvulus nervosus* Burm.f.	Convolvulaceae	Herb	A	Europe	M
48	*Corchorus aestuans* L.	Tiliaceae	Herb	A	Trop. America	Fu
49	*Crotalaria pallida* Ait	Fabaceae	Herb	A	Trop. America	Fi, Fu
50	*Crotalaria retusa* L.	Fabaceae	Herb	A	Trop. America	Fi, Fu
51	*Croton bonplandianum* Baill.	Euphorbiaceae	Herb	P	S. America	M
52	*Cucumis melo* L.	Cucurbitaceae	Climber	A	Iran and N. West	V

Sl no.	Plant species	Family	Life form	Habit	Nativity	Use
53	*Cuscuta reflexa* Roxb.	Cuscutaceae	Herb	A	Mediterranean	M
54	*Cynodon dactylon* (L.) Pers.	Poaceae	Herb	P	Trop. America	M
55	*Cyperus difformis* L.	Cyperaceae	Herb	P	Trop. America	M
56	*Cyperus iria* L.	Cyperaceae	Herb	P	Trop. America	M
57	*Datura innoxia* Mill.	Solanaceae	Shrub	P	Trop. America	M, Nox
58	*Datura metel* L.	Solanaceae	Shrub	P	Trop. America	M, Nox
59	*Delonix regia* (Boj.) Raf.	Caesalpiniaceae	Tree	P	Madagascar	O, Ave, Sf
60	*Duranta repens* L.	Verbenaceae	Shrub	P	America	O
61	*Echinochloa colona* (L.) Link	Poaceae	Herb	A	Trop. S. America	Fo
62	*Echinochloa crus-galli* Beauv.	Poaceae	Herb	A	Trop. S. America	Fo
63	*Eclipta prostrata* L.	Asteraceae	Herb	A	Trop. America (Bf1824)	M
64	*Eichhornia crassipes* (Mart.) Solm.	Pontederiaceae	Herb	P	Trop. America	St
65	*Emilia sonchifolia* (L.) DC.	Asteraceae	Herb	A	Trop. America	V, M
66	*Eucalyptus citriodora* Hook.	Myrtaceae	Tree	P	Australia	M, Fu, Sf
67	*Euphorbia heterophylla* auct. Non L.	Euphorbiaceae	Herb	A	Trop. America	O
68	*Euphorbia hirta* L.	Euphorbiaceae	Herb	A	Trop. America	M
69	*Euphorbia pulcherrima* Willd. ex. Klotz	Euphorbiaceae	Shrub	P	Mexico	O
70	*Euphorbia thymifolia* L.	Euphorbiaceae	Herb	P	Trop. America	NK
71	*Evolvulus nummularius* L.	Convolvulaceae	Herb	P	Trop. America	M
72	*Gnaphalium polycaulon* Pers.	Asteraceae	Herb	A	Trop. America	NK, Nox
73	*Gomphrena celosioides* Mart.	Amaranthaceae	Herb	A	S. America	Fo
74	*Gomphrena globosa* L.	Amaranthaceae	Herb	A	America	O
75	*Grangea maderaspatana* L.	Asteraceae	Herb	A	Trop. S. America	NK, Nox
76	*Grevillea robusta* Cunn. ex R.Br.	Proteaceae	Tree	P	Australia	Fu, Sf
77	*Helianthus annuus* L.	Asteraceae	Herb	A	America	O, Oil

Sl no.	Plant species	Family	Life form	Habit	Nativity	Use
78	*Heliotropium indicum* L.	Boraginaceae	Herb	A	S. America	M
79	*Hibiscus rosa-sinensis* L.	Malvaceae	Shrub	P	China	M, O
80	*Hyptis suaveolens* (L.) Poit.	Lamiaceae	Herb	A	Trop. America	M, Nox
81	*Impatiens balsamina* L.	Balsaminaceae	Herb	A	Trop. America	O
82	*Indigofera linnaei* Ali	Fabaceae	Herb	A	Trop. Africa	M
83	*Ipomoea batatas* (L.) Lam.	Convolvulaceae	Climber	P	Trop. America	Ft
84	*Ipomoea carnea* Jacq.	Convolvulaceae	Shrub	P	Trop. America	Fu
85	*Ipomoea hederifolia* L.	Convolvulaceae	Herb	A	Trop. America	M
86	*Ipomoea obscura* (L.) Ker-Gaw	Convolvulaceae	Herb	P	Trop. Africa	NK
87	*Ipomoea pes-tigridis* L.	Convolvulaceae	Herb	A	Trop. East Africa	M
88	*Ipomoea quamoclit* L.	Convolvulaceae	Herb	P	Trop. America	M
89	*Jatropha gossypifolia* L.	Euphorbiaceae	Shrub	P	Brazil	M
90	*Justicia gendarussa* Burm.f.	Acanthaceae	Shrub	P	China	M
91	*Kalanchoe pinnata* (Lam.) Pers.	Crassulaceae	Herb	P	Trop. Africa	O, M
92	*Kigelia pinnata* DC	Bignoniaceae	Tree	P	Africa	O, Ave, Sf
93	*Lagerstroemia indica* L.	Lytharaceae	Shrub	P	China	O
94	*Lantana camara* L.	Verbenaceae	Shrub	P	Trop. America	M, Nox
95	*Leonotis nepetifolia* (L.) R.Br.	Lamiaceae	Herb	A	Trop. Africa	M
96	*Leucaena leucocephala* (Lam.) de Wit.	Mimosaceae	Tree	P	Trop. America	Fu, Sf
97	*Ludwigia adscendens* (L.) Hara	Onagraceae	Herb	A	Trop. America	Sb
98	*Ludwigia octovalvis* (Jacq.) Raven	Onagraceae	Herb	A	Trop. America	M, Sb
99	*Ludwigia perennis* L.	Onagraceae	Herb	A	Trop. America	M, Sb
100	*Malvastrum coromandelianum* (L.) Garcke	Malvaceae	Herb	A	Trop. America	M, Fi
101	*Martynia annua* L.	Martyniaceae	Herb	A	Trop. America	M
102	*Mecardonia procumbens* (Mill.) Small	Verbenaceae	Shrub	A	T. N. America	NK
103	*Melochia corchorifolia* L.	Sterculiaceae	Herb	A	Trop. America	NK

Sl no.	Plant species	Family	Life form	Habit	Nativity	Use
104	*Merremia vitifolia* (Burm.f.) Hall.f.	Convolvulaceae	Herb	A	Trop. America	NK
105	*Mikania micrantha* L.	Asteraceae	Herb	A	Trop. America	Nox
106	*Mimosa pudica* L.	Mimosaceae	Herb	P	Brazil	M
107	*Mirabilis jalapa* L.	Nyctaginaceae	Herb	P	Peru	O, M
108	*Morus australis* Poir.	Moraceae	Tree	P	China	Ft
109	*Nicotiana tabacum* L.	Solanaceae	Herb	A	S. America	Nar, Sm
110	*Ocimum canum* Sims	Lamiaceae	Herb	A	Trop. America	M
111	*Opuntia stricta* (Haw.) Haw.	Cactaceae	Shrub	P	Trop. America	NK
112	*Oxalis corniculata* L.	Oxalidaceae	Herb	A	Europe	M
113	*Parthenium hysterophorus* L.	Asteraceae	Herb	A	Trop. America	Nox
114	*Passiflora foetida* L.	Passifloraceae	Climber	P	Trop. S. America	O, M
115	*Pedalium murex* L.	Pedaliaceae	Herb	A	Trop. America	M
116	*Peltophorum pterocarpum* (DC.) Backer ex K.Heyne	Caesalpiniaceae	Tree	P	Malaya	Ave, Sf
117	*Peperomia pellucida* (L.) Kunth	Piperaceae	Herb	A	Trop. S. America	M
118	*Peristrophe bicalyculata* (Retz.) Nees.	Acanthaceae	Herb	A	Trop. America	NK
119	*Phoenix sylvestris* (L.) Roxb.	Arecaceae	Tree	P	Trop. America	Ft, Fu
120	*Phyla nodiflora* (L.) Greene	Verbenaceae	Herb	A	Trop. America	M
121	*Physalis angulata* L.	Solanaceae	Herb	A	Trop. America	M, Ft
122	*Physalis minima* L.	Solanaceae	Herb	A	Trop. America	M, Ft
123	*Pistia stratiotes* L.	Araceae	Herb	P	Trop. America	M, St
124	*Pithecellobium dulce* (Roxb.) Benth.	Mimosaceae	Tree	P	Mexico	Ft
125	*Plumeria rubra* L.	Apocynaceae	Tree	P	S. America	O
126	*Portulaca oleracea* L.	Portulacaceae	Herb	A	Trop. S. America	M, V
127	*Portulaca quadrifida* L.	Portulacaceae	Herb	A	Trop. America	M, V
128	*Quisqualis indica* L.	Combretaceae	Climber	P	Malaya	O
129	*Rhoeo discolor* Hance.	Commelinaceae	Herb	P	Central America	O
130	*Richardia scabra* L.	Rubiaceae	Herb	A	S. America	NK
131	*Ricinus communis* L.	Euphorbiaceae	Shrub	P	Africa	M

Sl no.	Plant species	Family	Life form	Habit	Nativity	Use
132	*Ruellia tuberosa* L.	Acanthaceae	Herb	A	Trop. America	NK
133	*Saccharum spontaneum* L.	Poaceae	Herb	A	T. West Asia	Th, Fu, Fo
134	*Scoparia dulcis* L.	Scrophulariaceae	Herb	A	Trop. America	M
135	*Sida acuta* Burm.f.	Malvaceae	Herb	A	Trop. America	M
136	*Solanum nigrum* L.	Solanaceae	Herb	A	Trop. America	M
137	*Solanum torvum* Sw.	Solanaceae	Shrub	P	West Indies	M
138	*Sonchus asper* (L.) Hill	Asteraceae	Herb	A	Mediterranean	M
139	*Spathodea campanulata* Beauv.	Bignoniaceae	Tree	P	Trop. Africa	Ave, Sf
140	*Spermacoce articularis* L.	Rubiaceae	Herb	A	Trop. America	NK
141	*Sphaeranthus indicus* L.	Asteraceae	Herb	A	Trop. Africa	M
142	*Spilanthes acmella* (L.) L.	Asteraceae	Herb	A	North America	M
143	*Spinacia oleracea* L.	Chenopodiaceae	Herb	A	Europe	V
144	*Stachytarpheta jamaicensis* (L.) Vahl.	Verbenaceae	Herb	A	Trop. America	M
145	*Stylosanthes hamata* L.	Fabaceae	Herb	A	Trop. Africa	Fo
146	*Synedrella nodiflora* (L.) Gaertn.	Asteraceae	Herb	A	West Indies	M
147	*Tagetes erecta* L.	Asteraceae	Herb	A	Mexico	O, M
148	*Tagetes patula* L.	Asteraceae	Herb	A	Mexico	O, M
149	*Tamarindus indica* L.	Caesalpiniaceae	Tree	P	Trop. America	Ft
150	*Tecoma stans* (L.) Juss. ex Kunth	Bignoniaceae	Tree	P	America	O
151	*Thevetia peruviana* (Pers.) Merrill	Apocynaceae	Tree	P	Trop. America	M
152	*Thuja orientalis* L.	Cupressaceae	Tree	P	China	O
153	*Tribulus terrestris* L.	Zygophyllaceae	Herb	A	Trop. America	M
154	*Tridax procumbens* L.	Asteraceae	Herb	P	Mexico	M
155	*Triumfetta pentandra* A.Rich.	Tiliaceae	Herb	A	Trop. America	M
156	*Turnera ulmifolia* L.	Turneraceae	Herb	P	Trop. America	O
157	*Typha angustata* Bory & Chaub.	Typhaceae	Herb	P	Trop. America	Th, Fu, Fo
158	*Urena lobata* L.	Malvaceae	Herb	A	Trop. Africa	Fib, Fu
159	*Vernonia cinerea* L.	Asteraceae	Herb	A	S. America	M
160	*Waltheria indica* L.	Sterculiaceae	Herb	A	Trop. America	M

Sl no.	Plant species	Family	Life form	Habit	Nativity	Use
161	*Xanthium indicum* L.	Asteraceae	Herb	A	Trop. America	M, Nox
162	*Xanthium strumarium* L.	Asteraceae	Herb	A	Trop. America	M, Fu, Nox
163	*Zinnia elegans* Jacq.	Asteraceae	Herb	A	Mexico	O
164	*Zea mays* L.	Poaceae	Herb	A	America	F, Fu, Fo
165	*Ziziphus mauritiana* Lam.	Rhamnaceae	Tree	P	China	Ft

Note: F, food; FT, fruit; O, ornamental; not known; M, medicinal; Fu, fuel; Fib, fiber; V, vegetable; Sp, species; Nox, noxious; Sm, smoking; Co, compost; Sa, sacred plant; Sb, soil binder; Ch, chemical compounds; Ave, avenue; T, thatching; A, annual; P, perennial.

Table 1.
List of invasive alien plant species (IAPs) recorded from Rourkela Steel City, Sundargarh, Odisha.

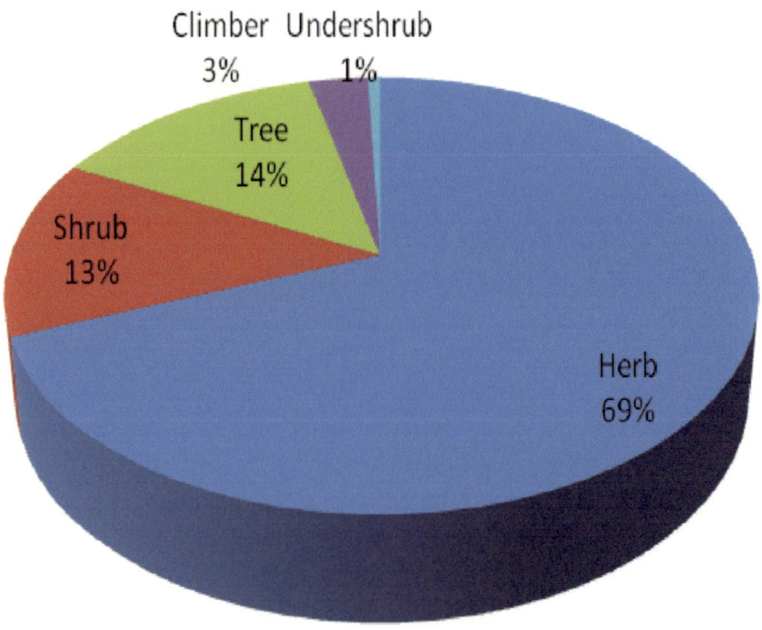

Figure 2.
Habit-wise distribution of invasive alien plant species in Rourkela.

Sl. no.	Habit	No. of species
1	Herb	114
2	Shrub	22
3	Tree	23
4	Climber	5
5	Undershrub	1

Table 2.
Habit of invasive alien plant species in Rourkela Steel City of Odisha.

America 77 species, Tropical South America 7 spp., Central America 1 spp., South America 10 spp., Tropical Africa 15 spp., Mexico 7 spp., and Europe 4 spp. IAPs are having negative impacts on the ecosystem and biodiversity of that region. Besides they are also found to be useful to local inhabitants. From the survey of literatures and interaction with local people about the IAPs, several plant species were used for different purposes like medicine, vegetables, fuels, fodders, etc. The study revealed that 87 spp. are used as medicine, while 18 spp. are used as fuel, 9 spp. used for fodders, and 30 spp. used for ornamental and avenue purposes. A total of 28 spp. were used as edible in the form of fruit, vegetables, oil, etc. Several species like *Argemone mexicana*, *Euphorbia hirta*, *Mimosa pudica*, *Ocimum canum*, *Calotropis* spp., *Croton bonplandianus*, *Catharanthus roseus*, etc. were mostly common medicinal plants used by local people, kabiraj and baidyas, while plants like *Cassia siamea*, *Leucaena leucophloea*, *Kigelia pinnata*, etc. were used for avenue plantation and social forestry. *Cannabis sativa* and *Nicotiana tabacum* were not only used for medicines, but also they are used for smoking as narcotic products. Alien species have been classified into naturalized and noxious species by various

Sl. no.	Nativity	No. of species
1	America	5
2	Central America	1
3	North America	2
4	South America	10
5	Tropical America	77
6	Tropical South America	7
7	Tropical North America	1
8	Africa	2
9	Tropical Africa	15
10	Tropical East Africa	1
11	Brazil	4
12	Australia	4
13	West Indies	5
14	Mexico	7
15	Malaya	2
16	Madagascar	2
17	Europe	4
18	China	6
19	Iran North West	1
20	Pantropic	1
21	Peru	1
22	South East Tropical Asia	1
23	Tropical West Asia	1
24	Central Asia	1
25	Mediterranean	4

Table 3.
Different geographic nativities of the invasive alien plants.

Sl. no.	Family	No. of species
1	Acanthaceae	3
2	Amaranthaceae	8
3	Apocynaceae	4
4	Araceae	1
5	Arecaceae	2
6	Asclepiadaceae	2
7	Asteraceae	24
8	Balsaminaceae	1
9	Bignoniaceae	3
10	Boraginaceae	1
11	Brassicaceae	1
12	Cactaceae	1
13	Caesalpiniaceae	11
14	Cannabinaceae	1
15	Capparaceae	4
16	Caricaceae	1
17	Casuarinaceae	1
18	Chenopodiaceae	2
19	Combretaceae	1
20	Commelinaceae	1
21	Convolvulaceae	9
22	Crassulaceae	1
23	Cucurbitaceae	1
24	Cupressaceae	1
25	Cuscutaceae	1
26	Cyperaceae	2
27	Euphorbiaceae	8
28	Fabaceae	5
29	Lamiaceae	3
30	Liliaceae	2
31	Lytharaceae	1
32	Malvaceae	5
33	Martyniaceae	1
34	Mimosaceae	4
35	Moraceae	1
36	Myrtaceae	1
37	Nyctaginaceae	2
38	Onagraceae	3
39	Oxalidaceae	1
40	Papaveraceae	1
41	Passifloraceae	1

Sl. no.	Family	No. of species
42	Pedaliaceae	1
43	Piperaceae	1
44	Poaceae	6
45	Polygonaceae	1
46	Pontederiaceae	1
47	Portulacaceae	2
48	Proteaceae	1
49	Rhamnaceae	1
50	Rubiaceae	2
51	Sapindaceae	1
52	Scrophulariaceae	1
53	Solanaceae	8
54	Sterculiaceae	2
55	Tiliaceae	2
56	Turneraceae	1
57	Typhaceae	1
58	Verbenaceae	5
59	Zygophyllaceae	1

Table 4.
Total number of IAPs' distributed family-wise species in Rourkela Steel City.

workers [8, 28, 39]. Many reports say different alien species become noxious after naturalized. *Ageratum conyzoides*, *Blumea lacera*, *Cassia alata*, *Lantana camara*, *Cassia tora*, *Parthenium hysterophorus*, *Xanthium* sp., *Datura* sp., *Cardamine scutata*, *Argemone Mexicana*, *Grangea maderaspatana*, *Hyptis suaveolens*, and *Gnaphalium polycaulon* were some noxious species found during the study. *Parthenium hysterophorus* was one of the highly noxious and abundantly grown plant species next to *Ageratum conyzoides* and *Lantana camera*.

From the taxonomical study, Asteraceae was the most dominant invasive family which dominated all other species due to its adaptive nature of seeds in different areas. The plant species have high reproductive potential to produce minute seeds so fast which disperse in new area through wind, air, and water. From the literature study, it was found that Asteraceae was more invasive in other areas of India [5, 7, 8, 23, 31, 32, 34, 35, 37] and also all over the world. Mallick et al. also found Asteraceae as the most dominating group of weeds among all other plant family groups. *Mikania micrantha*, a climber of Asteraceae, can climb trees and walls easily which inhibit the growth of the trees as well as cover the whole area so rapidly. *Parthenium hysterophorus* was another noxious plant of this family which could cause black fever disease. It grows very rapidly as its seeds disperse and grow so fast in new area which become invasive later. Annuals showed dominance over perennials among the invasive species as annuals complete life cycle and produce seeds to disperse in a short period in a year. Habit-wise distribution showed that herbaceous plants become more invasive than shrubs, climbers, and trees. Herbs have more tolerance to harsh condition and have great viability to grow in any condition which helps to become more invasive than others. Kumar et al. [11] found herbs as the more dominant plant group found in Rourkela flora.

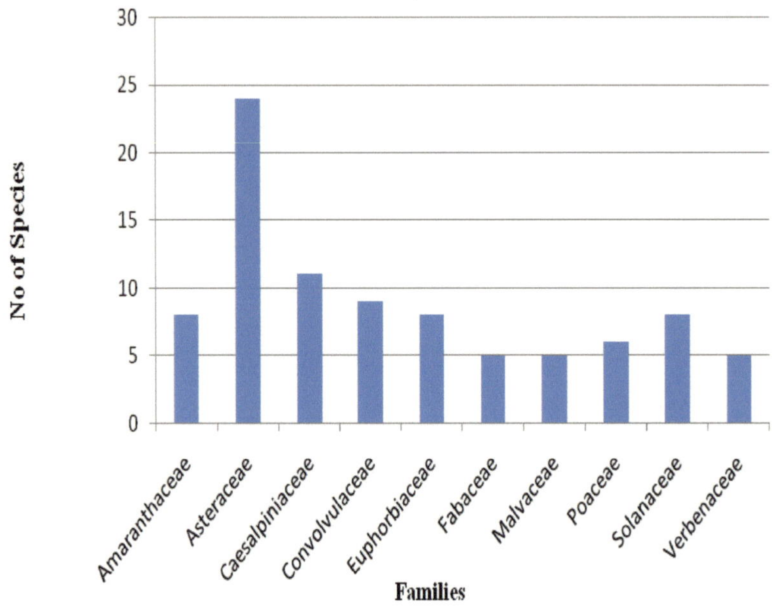

Figure 3.
Family-wise distribution of invasive alien plants in Rourkela Steel City of Odisha.

4. Conclusion

The invasive species are nonnative and exotic which are introduced intentionally for different purposes and sometimes accidentally introduced to a new area. The invasive species are more adapted to new areas by rapidly growing and reproducing more biomass than the native plant biodiversity. As a result, they can change the native ecosystem and become threats to the native ecosystem. IAPs also change the quality of soil, nutrient capacity, as well as the biodiversity present inside the soil. After invasion some invasive plants become narrow and noxious which affects the ecosystem with extinction of species and also affects the human health. Public involvement can be used and needed for early detection and reporting of infestations of the spread of new weeds as invasive species in the area. People should aware about the invasive species and its allelopathic effects on the environment and human health. Invasive species are now becoming more serious causing sustainable use of biodiversity and their impacts on invaded environment. Invasive alien plant species diversity in Rourkela of Sundargarh, Odisha, is a threat for the present flora due to their aggressive growth, colonizing ability, and adaptability. After invasion, their population growth increases rapidly in the new ecosystem; as a result they encroach crop fields, wastelands, and barren lands. The increased rate of invasion by alien species directly affects the agricultural economy and the biodiversity. Hence, eradication of IAPs should be done urgently. So awareness among local people is one of the methods to control IAPs. Besides this, the utilization of hidden medicinal potential can make IAPs beneficial to the people of the region. Moreover, the effect of IAPs in the economy, biodiversity, and human health is yet to be assessed. This study is based on diversity of invasive plant species found in different areas of Rourkela. Since the flora of Sundargarh district has not been beneficially and fully explored, this study will help in the compilation of flora of Sundargarh district and Rourkela in particular. Further studies reveal the allelopathic effects of IAPs on different plants, agricultural crops, and their ethnobotanical values.

Acknowledgements

Authors acknowledge the people of Rourkela for their kind cooperation and for sharing valuable information during the study.

Author details

Samarendra Narayan Mallick[1], Nirius Xenan Ekka[2], Sanjeet Kumar[3] and Sudam C. Sahu[4*]

1 Ravenshaw University, Cuttack, Odisha, India

2 School of Life Sciences, Sambalpur University, Sambalpur, Odisha, India

3 Ambika Prasad Research Foundation, Bhubaneswar, Odisha, India

4 Department of Botany, North Orissa University, Baripada, Odisha, India

*Address all correspondence to: sudam_rrl@yahoo.co.in

References

[1] Acharya BC, Khandagiri SN, Mallick SN, Maharana M. A survey of plant diversities in and around Rourkela. In: Proceedings of Env. Seminar, Govt. (Autonomous) College; Rourkela; 2007. pp. 08-24

[2] Acharya BC, Mallick SN, Dehury SS, Khandagiri SN. Survey of herbs in and around Rourkela. In: Proceedings of Env. Seminar, Govt. (Autonomous) College; Rourkela; 2008. pp. 35-54

[3] Acharya BC, Subudhi HN, Panda SP. Check list of economic plants of Rourkela and adjoining regions (Sundargarh district)—Odisha. Bulletin of Pure and Applied Sciences-Botany. 2010;**29**:53-57

[4] Das K, Duarah P. Invasive alien plant species in the roadside areas of Jorhat, Assam: Their harmful effects and beneficial uses. International Journal of Engineering Research and Applications. 2013;**3**(5):353-358

[5] Feng J, Zhu Y. Alien invasive plants in China: Risk assessment and spatial patterns. Biodiversity and Conservation. 2010;**19**:3489-3497

[6] Haines HH. Botany of Bihar and Orissa. Vol. 6. London: Arnold and Sons and West Nirman Ltd; 1921-1925

[7] Heywood V. Patterns, extents, and modes of invasions by terrestrial plants. In: Drake J et al., editors. Biological Invasions: A Global Perspective. New York: Wiley; 1989. pp. 31-60

[8] Huang QQ, Wu JM, Bai YY, Zhou L, Wang GX. Identifying the most noxious invasive plants in China: Role of geographical origin, life form and means of introduction. Biodiversity and Conservation. 2009;**18**:305-316

[9] Kull CA, Tassin J, Rangan H. Multifunctional, scrubby, and invasive forests? Wattles in the highlands of Madagascar. Mountain Research and Development. 2007;**27**:224-231

[10] Kumar P, Choudhury AK. Exotic species invasion threats to forests: A case study from the Betla national park, Palamu, Jharkhand, India. Tropical Plant Research. 2016;**3**(3):592-599

[11] Kumar S, Das G, Shin HS, Kumar P, Patra JK. Diversity of plant species in the steel city of Odisha, India: Ethnobotany and implications for conservation of urban bio-resources. Brazilian Archives of Biology and Technology. 2018;**61**:1-9

[12] Lal HS, Singh S, Kumar A, Mishra PK, Mishra K. Study of invasive and alien species in Jharkhand, India and its impact on environment. Journal of Ethnobiology and Traditional Medicine. 2012;**117**:167-177

[13] Mallick SN, Acharya BC. Phytodiversity survey of roadside herbs and their ethnomedicinal values in and around Rourkela Steel Township, Sundargarh, Odisha. Life Sciences Leaflets. 2013;**1**:46-55

[14] Mandal KK, Khora SS, Kar T. Aquatic angiosperm of Bonai Forest division, Sundargarh district, Odisha. Plant Science Research. 2017;**39**(1&2):12-18

[15] Maslo S. Preliminary list of invasive alien plant species (IAS) in Bosnia and Herzegovina. Herbologia. 2016;**16**(1):1-14

[16] Moktan S, Das AP. Diversity and distribution of invasive alien plants along the altitudinal gradient in Darjeeling Himalaya, India. Pleione. 2013;**7**(2):305-313

[17] Mooney HA, Hobbs RJ, editors. Invasive Species in a Changing World. Washington, D.C., USA: Island Press; 2000

[18] Naidu NT, Kumar OA, Venkaiah M. Invasive alien plant species in tropical forests of Eastern Ghats in northern Andhra Pradesh, India. Indian Forester. 2015;**141**(4):428-432

[19] Nayak SK, Satapathy KB. Diversity, uses and origin of invasive alien plants in Dhenkanal district of Odisha, India. International Research Journal of Biological Sciences. 2015;**4**(2):21-27

[20] Negi PS, Hajra PK. Alien flora of Doon valley, northwest Himalaya. Current Science. 2007;**92**(7):968-978

[21] Preston CD, Pearman DA, Hall AR. Archaeophytes in Britain. Botanical Journal of the Linnean Society. 2004;**145**:257-294

[22] Preston G, Williams L. Case study: The working for water programme: Threats and successes. Service Delivery Review. 2003;**2**(2):66-69

[23] Rao RR, Murugan R. Impact of exotic adventives weeds on native biodiversity in India: Implications for conservation. In: Rai LC, Gaur JP, editors. Invasive Alien Species and Biodiversity in India. Varanasi: Banaras Hindu University; 2006. pp. 93-109

[24] Rastogi J, Rawat DS, Chandra S. Diversity of invasive alien species in Pantnagar flora. Tropical Plant Research. 2015;**2**(3):282-287

[25] Reddy CS, Bagyanarayana G, Reddy KN, Raju VS. Invasive Alien Flora of India. USGS, USA: National Biological Information Infrastructure; 2008

[26] Reddy CS. Catalogue of invasive alien flora of India. Life Science Journal. 2008;**5**(2):85-87

[27] Reddy CS, Rangaswamy M, Pattanaik C, Jha CS. Invasion of alien species in wetlands of Samaspur bird sanctuary, Uttar Pradesh, India. Asian Journal of Water, Environment and Pollution. 2009;**6**(3):43-50

[28] Richardson DM, Pyšek P, Rejmánek M, Barbour MG, Dane Panetta F, West CJ. Naturalization and invasion of alien plants: Concepts and definitions. Diversity and Distributions. 2000;**6**(2):93-107

[29] Roder W, Dorji K, Wangdi K. Implications of white clover introduction in east Himalayan grasslands. Mountain Research and Development. 2007;**27**:268-273

[30] Saxena HO, Braham M. The Flora of Orissa. Vol. 4. Bhubaneswar, Odisha, India: Orissa Forest Development Corporation Ltd; 1994-1996

[31] Sekar K. Invasive alien plants of Indian Himalayan region—Diversity and implication. American Journal of Plant Sciences. 2012;**3**:177-184

[32] Sekar CK, Manikandan R, Srivastava SK. Invasive alien plants of Uttarakhand Himalaya. Proceedings of the National Academy of Sciences, India Section B: Biological Sciences. 2012;**82**(3):375-383

[33] Sharma GP, Singh JS, Raghubanshi AS. Plant invasions: Emerging trends and future implications. Current Science. 2005;**88**:726-734

[34] Sheikh DK, Dixit AK. Occurrence of invasive plant in three phytogeographical region of Bilaspur district of Chhattisgarh. Annals of Plant Sciences. 2017;**6**(12):1872-1878

[35] Singh KP, Shukla AN, Singh JS. State-level inventory of invasive alien plants, their source regions and use potential. Current Science. 2010;**99**(1):107-114

[36] Singh TB, Das AK, Singh PK. Study of alien and invasive flora of valley

district of Manipur and their control. International Journal of Innovative Research in Science, Engineering and Technology. 2015;**1**(2):616-626

[37] Srivastava S, Dvivedi A, Shykla RV. Invasive alien species of terrestrial vegetation of north eastern Uttar Pradesh. International Journal of Forest Research. 2014;**2014**:1-9

[38] Udaykumar M, Bharathidasan E, Sekar T. Invasive alien flora of Thiruvallur District, Tamil Nadu, India. Scholars Academic Journal of Biosciences. 2014;**2**(4):295-306

[39] Wu SH, Hsieh CF, Rejmanek M. Catalogue of the naturalized flora of Taiwan. Taiwania. 2004;**49**(1):16-31

Chapter 2

Invasive Species in the Amazon

Wanessa Almeida da Costa, Cinthya Elen Pereira de Lima, Sérgio Henrique Brabo de Sousa, Mozaniel Santana de Oliveira, Fernanda Wariss Figueiredo Bezerra, Jorddy Neves da Cruz, Sebastião Gomes Silva, Renato Macedo Cordeiro, Cintya Cordovil Rodrigues, Antônio Robson Batista de Carvalho, Priscila do Nascimento Bezerra, Pedro Alam de Araújo Sarges, Daniel Santiago Pereira, Antônio Pedro Silva de Souza Filho and Raul Nunes de Carvalho Junior

Abstract

One of the main reasons for environmental disturbances such as declination in pasture productivity and biodiversity losses is the high infestation of herbaceous weeds, generally referred to as "Juquira" in the Amazon region. If they are not adequately controlled, such infestation might lead to degradation of pasture, resulting in complete loss of productivity and subsequent abandonment of the area. In this sense, this chapter aims to describe the main invasive species present in the Amazon region, as well as to characterize both the old and innovative techniques of use in agriculture, in large and small scale, for the control of agricultural pests.

Keywords: Juquira, weed control, Amazon region, agriculture, infestation

1. Introduction

In the last years, large areas of forest in the Brazilian Amazon have been deforested for agricultural activities. Due to some changes in environmental conditions in these areas, some native plant species, which were originally restricted to the edges of forests or clearings, increased in abundance. As a consequence, these species ecological and agronomic importance has changed significantly. For example, many have become important weeds in active farming or have thrived in abandoned areas [1]. In view of this, vegetative invasions are among the main causes of biodiversity loss. In this context, environments subject to anthropic disturbances, such as the Amazon region, are considered more prone to propagation and invasion of species that cause degradation and biological variety loss [2–4].

Weeds interfere with agricultural crops by reducing mainly the yield. This interference occurs due to the competition for water, light, nutrients, and chemical inhibition, affecting the germination and development of cultivated plants. Indirectly, weed species can cause damage to crops by harboring insect pests, fungi, and nematodes and make it difficult to harvest and also depreciate the harvested product quality [5].

According to Lorenzi [6], the weed conception is relative as no plant is exclusively harmful. On the other hand, cultivated plants correspond to those species sown or cultivated by man, while wild plants are all plant species that are born and reproduce spontaneously, which may interfere with crop production, domestic animals welfare, and the view aspect where they occur [6, 7]. Invasive plants are defined as any plants, either cultivated or wild, that vegetate in places where their presence is not desired [7].

Weeds have emerged from a dynamic process of evolution by adapting itself to environmental disturbances caused by nature or man through agriculture. This evolution continues until today in response to the agriculture modernization. Among the modern techniques used in agriculture, the herbicides used to control weeds have provided a fairly rapid evolution, making them in some situations resistant to these chemicals. The evolution of weed population resistant to herbicides is a growing problem in many countries. Thus, many research works are conducted, especially in the last 20 years, in order to study distribution, resistance mechanisms, genetics, and management of these populations [8].

According to da Silva et al. [9], in the cultivated pasture areas, the invasive plants correspond to the main maintenance cost factor. However, the relevance of their role in ecosystems and the available information on the biology and management of these plants is limited, with the majority of them being floristic or relating to the elaboration of control strategies, such as those mentioned in Dias Filho [1], Mascarenhas and Dutra [10], and Dutra et al. [11] work. Studies show that there are innumerable direct and indirect effects of invasive plants on man's agricultural activities, ranging from competition for essential mineral nutrients, light, water, and space, to alternative lodging of pests, nematodes, and parasitic plants [6, 12, 13].

In this sense, the objective of this work was to characterize some of the most commonly found invasive species in the Amazon, as well as to describe the main methods of these weed control.

2. Main invasive species in the Amazon region

2.1 Pau-de-lacre (*Vismia guianensis* (Aubl.) Choisy)

For invasive plants, successful invasion may be related to superiority in competition with native species. In this scenario, persistent perennial species are the ones that cause the most damages and live for several years and, in most cases, reproduce both by seeds (sexed) and vegetative (asexual) [14]. In this category, there are the most problematic species for agricultural pastures, such as lacre (*Vismia guianensis*) (**Figure 1**), which are inconvenient mainly for the extensive or semi-intensive production systems [15, 16]. *Vismia guianensis* is a plant, which belongs to the family Hypericaceae, order Malpighiales, and class Magnoliopsida, and it is distributed in North and Northeast Brazil [15, 16].

Vismia guianensis species, commonly known in the north of Brazil by the name of "pau-de-lacre," is a rough perennial plant with a brown coloration [18]. The leaves are greenish and shiny, and on the dorsal part (bottom), they are rough and ferruginous in color. When the leaves are removed, an orange colored viscous liquid flows from

Figure 1.
Species lacre (Vismia guianensis), adapted from [17].

the cut, and the inflorescences of this species present yellow flowers with globally green fruits, containing a large number of seeds. *V. guianensis* species is a bushy plant, distributed in the forest edge, being able to reach from 2 to 5 m height [19]. The flowering extends from November to March, extending until May; in some individuals, the peak of flowering occurs between December and January. The fruits conserve color even when mature, thus visually impossible to distinguish from the immature [14, 16, 17].

V. guianensis propagates aggressively from the stems and roots. This process, according to [19], is the main stimulus to growth after the cutting or burning of the species propagation areas. And one of the main problems arising from the *V. guianensis* dispersion is the competition for environmental factors (light, water, nutrients, and space) with native or cultivated plants in the environment [20]. This species also has high growth, proliferation, and dispersal capacity and is capable of modifying the composition, structure, or ecosystem function.

2.2 Capim-navalha (*Paspalum virgatum* L. (Gramineae))

Another invasive species frequently found in agricultural production pastures is the *Paspalum virgatum*, commonly known as "capim-navalha" (**Figure 2**). It is an invading pasture grass, recognized as a weed in the Amazon. Its propagation leads to diseases that cause progressive death of susceptible grasses, opening space for the colonization of weeds and leading to pasture degradation [15]. *Paspalum virgatum* is a plant species, which belongs to the family Poaceae, order Gramineae, Monocotyledons class, with hypogynic stamens. It is a perennial, herbaceous, erect weed, with a size of up to 1.50 m height, has a great protein value, but when adult, it becomes fibrous and not palatable to animals. Its flowering occurs between the months of October and May, season of greater intensity of rains, especially in Amazonian regions. Plants of this species usually inhabit humid environments, borders of streams, rivers and are also frequently found in pastures [15].

Capim-navalha (*P. virgatum*) is a Central America and South America native species [21]. In Brazil, it is found in all states of the North, especially in humid areas with high multiplication capacity. Capim-navalha also receives other denominations, such as navalhão, capim-duro, capim-cabeçudo, capim-taripucu, and capim-capivara [15]. *P. virgatum* is a cespitosa and rhizomatous grass, with clumps reaching 1.5 m in height, fibrous roots and upright leaves with 50 to 75 cm long and 1 to 2 cm width [22].

Figure 2.
Capim-navalha (Paspalum virgatum), adapted from [15].

The problems caused by their high multiplication capacity compete with fodder (common name given to feeding or lining the place where the animal sleeps), especially in moist soils where cattle graze only young plants of capim-navalha, while the equines appreciate the seeds and help to spread the infestation in the pasture. It also interferes negatively on plant growth through competition for water, nutrients, and allelopathy, with influence on pasture establishment and pasture regrowth ability after grazing and control difficulty by conventional methods [9, 15].

2.3 Malícia (*Mimosa pudica* L.)

Mimosa pudica L (**Figure 3**) is an invasive plant species, belonging to the genus *Mimosa*, subfamily Mimosoideae, family Fabaceae or Leguminosae and order Fabales [23]. *M. pudica* L. is a semi-prostrate herb, also very found in the Amazon, prickly or underbush up to 0.5 m height [24], of branched stems, with numerous bristly and deflected hairs [25]. The leaves are very sensitive, folding when touched

Figure 3.
Malícia (Mimosa pudica L.)

and digitally dividing with one or two pairs of sessile, alternates, petiolate, stipulated and linear lanceolate hairs [24, 25]; the flowers on the globular head are pink with prickly peduncles, while the fruits are indehiscent, simple, dry, with 1–1.6 cm in length and 0.4–0.5 cm in width, housing two to five seeds [24].

M. pudica is known with several colloquial names. The most recurrent are: malícia [26], planta tímida, planta sensível, and planta humilde [24, 27, 28]. This invasive plant species belongs to the Fabaceae family [29, 30], showing the following variations: *Mimosa pudica* var. hispida Brenan, *Mimosa pudica* var. tetrandra (Humb. & Bonpl. ex Willd.) DC., and *Mimosa pudica var. unijuga* (Walp. & Duchass.) Griseb [29].

This species is native to Africa and Asia, but it is common to be found in North and South America and with excellent adaptation in Brazil, especially in the Amazon region [29, 31]. According to Azmi et al. [25], perhaps it is native to many or all the tropics of the New World, and today its distribution may be pantropical.

It grows more in well drained soils, but also grows in scalded or eroded soils, soils with low concentrations of nutrients, at sea level or at altitude up to 1300 m, intolerant to shade, not competing with large vegetation, or growing under the canopy of trees [25]. This species is an invasive plant of pastures, agricultural areas, orchards, roadsides and roads, cut areas, areas disturbed by construction, polluted areas, among other productive areas of commercial and/or open fields [24, 25, 30].

The adaptation and proliferation of this weed is due to factors such as large seed production and anti-herbivore defenses (this occurs with the leaves rapid movement and the petiole decline) [25, 32–34]. These leaves are also in response to stressors such as electrostimulation, wound, wind, vibration, touch, drought, change of lighting, and warm or cold stimuli, which help the plant to protect itself or adapt to a particular environment condition where the vegetable is inserted [25, 27], such as at light levels [25, 27, 28]. In addition, in the plant radicle nodules occurs a symbiotic bacterial association that transforms the atmospheric nitrogen in a useful way for the plant, benefiting its development [30]. This species also has cylindrical roots with a slightly rough surface or wrinkled longitudinally, tapered, with secondary and tertiary ramifications, varying in length and thick up to 2 cm [24]. They are still capable of producing carbon disulfide, which selectively inhibits the rhizosphere colonization by mycorrhizal and pathogenic fungi [25]; and finally, the soils, which are often burned, allow the spread of this weed [27].

This plant produces the amino acid mimosine and its metabolite, 3-hydroxy-4-(1H)-pyridone (DHP) which, when ingested, is toxic to horses, cattle, pigs, and sheep, causing hair loss, low growth, oral ulcerations, and goiter not prevented by iodine supplementation [35].

2.4 Mata-pasto (*Senna obtusifolia* (L.) Irwin & Barneby)

Another important invasive species is the *Senna obtusifolia* (**Figure 4**). This is an invasive and erect stem plant without spines, measuring from 1.5 to 2.0 m in height, and the leaves are in pairs with three pairs of leaflets, hairless, and not brittle [36]. The name *Senna obtusifolia* comes from Latin *obtus* (opaque or blind) and *fólio* (leaf) [37]. This species of invasive plant is very common, infesting crops in tropical and subtropical regions of the world. It is an annual plant with woody base, belonging to the Fabaceae family, subfamily Caesalpinioideae, and order Fabales that reproduces itself by seeds, that are in the form of a cluster with yellow petals and sprout especially in spring and summer [38–40]. In the same way as other invasive species, *Senna obtusifolia* produces seeds on a large scale [41]. In the case of this species, this is due to the fruits with multiple seeds [38]. It also has an ultra-aggressive radicle system, giving it a high competitive capacity, even in periods when the soil has low

Figure 4.
Mata-pasto (Senna obtusifolia).

hydric availability [39], although it does not present nitrogen fixing nodules in the roots, which is common in many species of *Senna* and that is of extreme importance for invasive plants [41].

Researchers believe that *Senna obtusifolia* originates in the Caribbean and in tropical South America [40], but has spread widely and exhibits a global pantropical distribution [39, 40], as it can be found in Africa, India, Sri Lanka, Pakistan, Central America, Malaysia, Philippines, Indonesia, Papua New Guinea, South America, Caribbean, USA, and Australia [37, 40]. This species is present also in environments of 1600 m altitude, as in Mexico and Tanzania [40].

It is known by many different names in different parts of the world, but the most common are: mata-pasto [26, 36], sicklepod [39], fedegoso, and Feijão-Java [37, 40]. It is a very aggressive weed of agricultural areas of a wide variety of crops and in several countries, being predominant in plantations of soybean, peanuts, cotton, sugarcane, corn, disturbed areas, such as animal husbandry pastures, and open ecosystems [37–40, 42].

Weed competition and interference affect agricultural productivity, thereby significantly reducing the productivity of the planted crop, as well as altering the structure and function of the local natural ecosystem [40]. Another prejudice caused by *Senna obtusifolia* occurs when bovine animals consume the green leaves and fruits in pastures and/or feed contaminated with leaves, stems, and seeds of the plant, causing serious poisoning, which can lead the animal to death [36].

2.5 Tiririca (*Cyperus rotundus* L.)

Cyperus rotundus L. species (**Figure 5**), also known as tiririca or erva-cidreira, belongs to the Cyperaceae family. This is the third largest family of monocotyledonous plants [43]. It is a colonial herb, perennial, has 7–40 cm of height with fibrous roots, and reproduces largely by rhizomes and tubers. Rhizomes can grow in any direction on the ground, those growing up produce shoots and roots; the rhizomes that grow down horizontally form individual tubers or tuber chains. Mature individual tubers are reddish brown, about 12 mm thick and ranging between 10 and 35 mm in length. Tuberous roots act as the main dispersion units over time, remaining dormant in the soil for long periods. Tuber dormancy causes irregular emergence, contributing to the persistence of the propagules of this species [44].

Figure 5.
Tiririca (Cyperus rotundus).

The leaves are dark green, bright, narrow, and similar to herbs, ranging from 5 to 12 mm wide, and 50 cm long. The vertical stems support a branched inflorescence with bisexual flowers with three stamens, and a pistil with three stigmas. Nuts are rarely produced [44].

Cyperus rotundus is a weedy plant that is difficult to handle and causes damages in several commercial crops. Damage results from competition throughout the cycle, but the most critical periods are in the early stages of crop development and crop reforms. For it being a perennial species, and for its broad adaptability to many agricultural environments and the ability to reproduce sexually and asexually, *C. rotundus* is among the 20 most damaging species in the world [45].

On the other hand, *Cyperus rotundus* has its medicinal imprint. It is widespread in many tropical and subtropical regions of the world [46] and is considered to have originated in India for over 2000 years and is regarded as one of the best herbs for medicinal purposes. Studies indicate that the rhizomes of *C. rotundus* are used as traditional remedies for the treatment of stomach and intestinal disorders and inflammatory diseases in Asian countries [47–49]. Studies on the ethnobotanical use of *C. rotundus* showed that rhizomes were used to treat diseases of aging, apoptosis, atherosclerosis, cancer, cystitis, epilepsy, genotoxicity, hirsutism, nociception, and prostatitis [50]. It is reported that the tuberous part of *C. rotundus* is used for the treatment of dysmenorrhea and menstrual irregularities since antiquity [51].

2.6 Dente-de-leão (*Taraxacum officinale* L. Weber ex FH Wigg)

The species *Taraxacum officinale* (dente-de-leão) (**Figure 6**) is a perennial herb native of Europe, considered an aggressive invasive species worldwide [52]. In its

Figure 6.
Planta dente-de-leão (Taraxacum officinale), adapted from [23].

native distribution, *T. officinale* is present in alpine environments, mainly restricted to disturbed sites [53].

Widely distributed in the northern hemisphere, the *Taraxacum* genus is a member of the Asteraceae family, of Cichorioideae subfamily. *Taraxacum officinale* (dente-de-leão) is a perennial stemless weed, green leaves are grouped at the plant base, and the whole herb contains white latex. The flowering stems stand out with yellow flowers. Dente-de-leão plant is deeply ingrained, which means that the plant is also capable of producing a new plant even after its aerial part has been clearly cut. The herb is harvested between spring and autumn when the plant begins to bloom. Whole herbs are cleaned and dried in the sun until their moisture content is less than 13.0% [54].

Taraxacum officinale shows high tolerance to abiotic stress and efficient use of resources due to high plasticity in morphological and physiological characteristics [55–57]. Thus, when it presents favorable abiotic conditions, *T. officinale* shows greater abundance, physiological performance, accumulation of biomass, survival, and seed production [58, 59].

Taraxacum officinale has already been recognized as a useful passive bioindicator for heavy metals in urban areas [60], as well as a potential indicator for several trace elements, but only in highly polluted industrial areas. In response to vestigial elements, this species exhibits some micromorphological alterations. However, the lack of visual effects and the occurrence in industrial areas of medium pollution with reduction of heavy metals content in soils may indicate their potential for bioindication and phytoextraction [61]. It has all the necessary resources for good bioindicators: it is widely spread geographically, characterized by relatively high tolerance to environmental pollutants, and shows a correlation between the pollution level of a certain environment element (air and soil) and these substances concentration (metals heavy, polycyclic aromatic hydrocarbons) in plant tissues [62, 63].

3. Traditional methods of weed control

The invasive plants control has great relevance in agriculture, since these species bring losses to native species, communities, and ecosystems with the loss of their nutrients, decrease in yield, and quality of the crop, bringing direct impacts to human life and other species. Several techniques can be used to reverse, interrupt, or decelerate infested areas making them healthy again.

There are methods used to reduce the development and performance of invasive plants. The three most viewed are: (a) prevention, which involves preventive measures to introduce these plants into an ecosystem, (b) eradication, which contemplates the extermination, including its seeds however studies show that it is practically impossible to be carried out in large areas and it is economically unviable, and (c) control, that according to Tu et al. [64], some of the more traditional control options are manual, mechanical, competition between native plants, grazing, herbicides, prescribed fire, solarization, and flooding involving chemical, physical, and biological methods [65, 66].

3.1 Chemical methods

The chemical control main advantages are: (a) efficiency; avoids the competition of weeds since the crop implantation; (b) allows controlling weeds in rainy season, when mechanical control is impracticable; (c) does not cause damage to the crop roots; (d) does not revolve the soil; (e) allows better distribution of the economic crop plants in the area; (f) controls weeds in the main crop line; (g) and is of rapid operation. While important for reducing costs and increasing productivity, its indiscriminate use is a global environmental problem as it affects living organisms. Among the disadvantages are the cost, generally higher than the other methods; requires adequate equipment; may be toxic to man and animals; contaminates the environment and can leave residues in soil and food [67, 68]. The intensive use of phenoxyalkanoic acid herbicides in agriculture has an adverse effect on the environment that involves water pollution, among other phenomena. In many countries, phenoxyalkanoic acid herbicides have been found in groundwater, surface, and potable water in concentrations that have exceeded the maximum permissible limits, which determines an environmental problem [69].

Chemical control is important mainly in places where there is high weed infestation and low availability of water and nutrients, and the time available for control is reduced due to the area size or the lack of high-performance equipment. In large soybean plantations, chemical control is the most commonly used method due to agility and efficiency. Farmers using the chemical method should be aware of the interactions between the variety being used and the herbicide to be applied, as some cultivars are more sensitive than others to certain herbicides [68].

It is possible to halve the amount of herbicides without loss in weed control effectiveness and crop yield by combining chemical weed control in line with cross-line collection [70].

3.2 Physical methods

The physical control begins with several aspects that range from the use of a suitable crop to the chosen place, study of planting season, and adequate seeds, to the study of characteristics such as configurations, density, soil, and climate. Another physical control occurs with the grinding and pre-incorporation of fertilizers and remaining plants with subsequent plowing in moist soil, as this can considerably reduce the growth of invasive plants. Still evaluating the physical methods, there is a third alternative that is the use of crop rotation, which reduces the incidence not only of invasive plants but also of pests and diseases. And, as a last indication, the mechanic, who uses from hoes, passing through tractors, or even animal traction. This method was widely used for its low cost, and the need of not very modern equipment, so that its range is still great among rural producers and families to contain invasive plants and adequate development of their planting [71].

3.3 Biological methods

Biological methods are most appreciated in weed control against chemical and physical methods due to the viability of use in any situation. According to Weed Science Society of America [72], the biological control of weeds is defined as "the use of an agent, a complex of agents, or biological processes to bring about weed suppression"; some examples of agents used are arthropods (insects and mites), plant pathogens (fungi, bacteria, viruses, and nematodes), fish, birds, and other animals. Some advantages in using biological methods compared to other methods are related to the decreased risk of soil, water, and food contamination by herbicide residues, bringing healthier and sustainable cropping systems, besides being low cost and self-sufficient [73].

4. Recent methods of weed control

4.1 Use of agrochemicals

The predominant agriculture in the world, in addition to high productivity, is also characterized by its dependence on fossil energy sources, such as fertilizers and pesticides. However, the increasing increase in the use of agrochemicals as it has been happening may not be sustainable over time, not only because these products pollute the environment and promote the intoxication of animals and humans but also because new breeds of insects and new species of invasive plants, both resistant to insecticides are appearing with increasing frequency [74].

The number of cases of resistance to insecticides and fungicides increased rapidly after the 1950s and 1960s [75]. Since the first report by Ryan [76], which observed resistant *Senecio vulgaris* biotypes to the herbicides belonging to the chemical group of the triazines, it have been observed an increasing number of weed species with biotypes resistant not only to triazines but also to other classes of herbicides.

Weed resistance to herbicides may result from biochemical, physiological, morphological, or phenological changes of certain weed biotypes. Many cases of resistance to herbicides result from either altering the herbicide site of action or increasing its metabolism, or the departmentalization and compartmentalization of the herbicide in the plant. Although these general mechanisms are similar to some crop selectivity mechanisms, which allow them to survive herbicide exposure, specific herbicide resistance mechanisms in weeds typically differ substantially from those responsible for crop selectivity [77].

Some natural chemicals are used as a model for obtaining new herbicides. In addition, chemicals with proven allelopathic activity can be concentrated and have their allelopathic effect potentiated in the laboratory [78].

4.2 Alternatives in agricultural pest control

Over the past seven decades, considerable efforts have been expended to detect plant species with potential for use in a variety of human activities, such as medicine, cosmetics, hygiene, and food industry. Obviously, popular knowledge and medicinal use of many plant species, especially by indigenous communities, for example, played a prominent role at the beginning of the research, allowing studies based on a minimum of available information, which reduced research time and speed advances [79].

Theoretically, all plants are capable of producing chemically highly diversified compounds, some of which have potential for use in weed management. This specificity can be detected in both native and cultivated plants, although in the first—even because they have not been domesticated—such properties may be more auspicious. Identifying and selecting, depending on the degree of toxicity, plants species with potential for such purposes have become a primary activity in many universities' laboratories and research institutes around the world. Good examples are found in the Souza Filho et al. and Iqbal et al. work [80, 81].

Studies in the literature have shown that certain plants in nature have the ability to synthesize compounds that can act in the development and growth of other organisms in the same ecosystem [82]. For Miller [83], the study of the allelochemicals is fundamental, because the action of such substances is of extreme importance for the understanding of the organism interactions in both natural and agricultural ecosystems.

In addition to these aspects, the perception that these plants may provide new prospects for agricultural exploitation adds to the innumerable possibilities of using biodiversity, not only for microorganisms but also for plant species [79].

4.3 Allelopathy: a natural method for weed control

Müller [84] proposed the term interference to classify the different types of change that develop among the various components of a plant community. This term was subdivided by Szczepahiski [85] in three groups: allelospoly, allelopathy, and allelomediation. Allelospoly or competition was defined as interference caused by the different components of the ecosystem by removing from the environment elements such as water, nutrients, and light, lowering it to levels that hinder normal development of others. Allelopathy is the alteration caused by the release of a chemical substance, elaborated by one or more components that affect certain elements of the community, and allelomediation or indirect interference is defined as the effects that alter the physical or biological environment, with reflexes in the living beings.

Allelopathy is a phenomenon that occurs largely in nature and has been postulated as one of the mechanisms by which some plants may interfere with others in their neighborhoods, changing the pattern and density of vegetation in a plant community [82].

From an agronomic point of view, allelopathy is of great interest as it allows not only the selection of pasture plants that can exert a certain level of control of some undesirable species such as invasive plants but also the establishment of grass species and forage legumes that are not strongly allelopathic among them and that can thus compose more balanced pastures, with favorable effects on productivity and longevity [86].

The most frequent studies on allelopathy are related to the effects of plant extracts on the germination and growth of others. In general, germination is less sensitive to allelochemicals than seedling growth [87].

Allelopathy may play an important ecological role in the near future as a source of new chemical substances with possibilities of use in Brazilian agriculture, similar to what already occurs in other countries such as Japan, Germany, and the United States, as a pasture management tool, and/or as supplier of basic structures for agricultural biodefensive production [74].

Chemicals that impose allelopathic influence are called allelochemicals. Allelochemicals have a very diverse chemical nature ranging from simple hydrocarbons to complex polycyclic compounds with high molecular weight. Such compounds are in general short chain fatty acids, essential oils, diterpenes, alkaloids, steroids, phenolic compounds: flavonoids, naphthoquinones, anthraquinones, and coumarin derivatives [88].

Numerous chemical substances with allelopathic potential are described in the literature, but some chemical classes deserve greater attention [89]. The saponins may be formed by glycosylated triterpenoids with a hydrophilic polysaccharide chain or by hydrophobic steroids, which give them the detergent property and, consequently, the ability to bind to cell membranes, affecting cellular functioning. They are known for their hemolytic properties and toxicity to molluscs, insects, and fungi [90].

Flavonoids are present in plants in various forms and with varied functions. They include flavonoids, flavones, flavanones, catechins, anthocyanins, proanthocyanidins, and isoflavonoids, among others. In addition to the pigment functions, attractive or repellent of herbivores, protection against UV radiation, these substances have allelopathic effects, being able to inhibit the growth of plants and fungi [82, 91, 92].

Finally, the alkaloids encompass more than 12,000 structures already described, behind only the terpenoids. Approximately 20% of plant species accumulate alkaloids, molecules characterized by low molecular weight, and origin from phenylalanine, tyrosine, tryptophan, and lysine [88, 93, 94]. According to Rice [82], several alkaloids are able to inhibit the growth of bacteria, in addition to being toxic to some invertebrates.

Most studies report that allelopathic compounds act as inhibitors of germination and growth [95]. However, some studies have demonstrated that these compounds can also act as growth promoters [96, 97]. Apparently, most if not all organic compounds that are inhibitory in some concentrations are stimulants in lower concentrations [82, 88].

Invasive species	**Control methods**			**Site**
	Biological	**Chemical**	**Physical**	
Hypericum perforatum	Herbivorous insects (classic)	—	—	Southern and Western parts of Australia
Cryptostegia grandiflora	Rust of *Maravalia cryptostegiae* (modern)	Herbicides (modern)	—	Tropical Queensland, Australia
Spartina alterniflora	—	Herbicides (modern)	Cutting and crushing (classic)	Willapa Bay
Euphorbia esula	*Spurgia capitigena* (modern)	—	—	North America
Cirsium arvense	*Aceria anthocoptes* (modern)	—	—	United States of America
Schinus terebinthifoliu	*Crasimorpha infuscata* Hodges	—	—	Hawaii—United States of America
Acacia dealbata	—	Herbicides (modern)	Cutting and pruning	Northwest Spain
Pteridium aquilinum	—	Herbicides (modern)	Firing, manual and/ or mechanical removal	Northern California
Bromus tectorum	—	—	Cutting and defoliation	Eastern Oregon (United States of America)
Centaurea maculosa	Biocontrol (modern)	—	Firing	North America, Western United States of America
Chondrilla juncea	*Aceria chondrillae* (modern)	—	—	Canada

Table 1.
Weed control methods in different parts of the world [101–109].

Many of these studies approach, under laboratory conditions, the effects of aqueous or even hydroalcoholic crude extracts on seed germination and elongation of the radicle and hypocotyl of different weed species [96, 98, 99].

It is known that under field conditions, there is no way to separate the effects attributed to competition from those of allelopathy, since the influence of plants on other species in their neighborhood is a complex combination of competition interference and allelopathic chemical reactions [100].

Table 1 shows the main weed control methods that are applied in the world.

5. Conclusion

The weeds are the main bioeconomic factor to impose limitations to the agricultural activities' performance developed in the tropical regions, such as Amazon, especially in the quality and productivity of the pastures offered to grazing animals, which are severely affected by these types of plants. Therefore, its adequate control is of fundamental importance for the most varied aspects, such as profitability, agronomic performance, and activities longevity. Another important aspect is the herbicides indiscriminate use reduction, which, as a consequence reduces the environmental and human health damages. Finally, the search for alternatives to the herbicides on the market, such as the numerous metabolites produced by plants, can provide surprising diversity of chemical structures, which offer excellent prospects for increasing the search for more specific types of herbicides and less damaging than those in use.

Author details

Wanessa Almeida da Costa[1,2*], Cinthya Elen Pereira de Lima[5], Sérgio Henrique Brabo de Sousa[2], Mozaniel Santana de Oliveira[2], Fernanda Wariss Figueiredo Bezerra[2], Jorddy Neves da Cruz[3], Sebastião Gomes Silva[3], Renato Macedo Cordeiro[1,2], Cintya Cordovil Rodrigues[1,2], Antônio Robson Batista de Carvalho[2], Priscila do Nascimento Bezerra[2], Pedro Alam de Araújo Sarges[5], Daniel Santiago Pereira[4], Antônio Pedro Silva de Souza Filho[4] and Raul Nunes de Carvalho Junior[1,2]

1 Program of Post-Graduation in Natural Resources Engineering (PRODERNA/ ITEC), Federal University of Pará, Belém, Pará, Brazil

2 LABEX/FEA (Faculty of Food Engineering), Federal University of Pará, Belém, Pará, Brazil

3 Program of Post-Graduation in Chemistry, Federal University of Pará, Belém, Pará, Brazil

4 Embrapa Western Amazon, Belém, Pará, Brazil

5 Program of Post-Graduation in Mechanical Engineering, Federal University of Pará, Belém, Pará, Brazil

*Address all correspondence to: wanessa.almeida712@yahoo.com.br

References

[1] Dias Filho MB. Plantas invasoras em pastagens cultivadas da Amazônia: Estratégias de manejo e controle. 1st ed. Belém: EMBRAPA-CPATU; 1990. Epub ahead of print 1990. DOI: 10.13140/RG.2.1.3427.4403

[2] Traveset A, Richardson DM. Biological invasions as disruptors of plant reproductive mutualisms. Trends in Ecology & Evolution. 2006;**21**:208-216

[3] Simberloff D, Nuñez MA, Ledgard NJ, et al. Spread and impact of introduced conifers in South America: Lessons from other southern hemisphere regions. Austral Ecology. 2010;**35**:489-504

[4] Petenon D, Pivello VR. Plantas invasoras: Representatividade da pesquisa dos países tropicais no contexto mundial. Nature Conservation. 2008;**6**:65-77

[5] Brighenti AM. Manual de Identificação e Manejo de Plantas Daninhas em Cultivos de Cana-de-açúcar. 1st ed. Embrapa Gado de Leite: Juiz de Fora; 2010

[6] Lorenzi H. Plantas Daninhas do Brasil: Terrestres, Aquáticas, Parasitas e Tóxicas. 4th ed. Instituto Plantarum: Nova Odessa; 2008

[7] Silva AF, Leitão Filho HF. Composição florística e estrutura de um trecho de mata atlântica de encosta no município de Ubatuba (São Paulo, Brasil). Revista Brasileira de Botanica. 1982;**5**:43-52

[8] Christoffoleti PJ, Victoria R, Filho S, et al. Resistência de plantas daninhas aos herbicidas. Planta Daninha. 1994;**12**:13-20

[9] da Silva MF, Gurge ESC, Filho AP d SS, et al. Leguminosae invasive species in cultures in the northeast of Pará, Brazil. Boletim do Museu Paraense Emílio Goeldi. Ciências Humanas. 2013;**8**:63-74

[10] Mascarenhas REB, Dutra S. Plantas daninhas de uma pastagem cultivada de baixa produtividade no nordeste paraense. Planta Daninha. 1999;**17**:1999

[11] Dutra S, Souza Filho APS, Mascarenhas REB, et al. Controle Integrado de Plantas Invasoras em Pastagens Cultivadas no Município de Terra Alta, Nordeste Paraense. 25th ed. Belém: Embrapa Amazônia Oriental (Boletim de Pesquisa e Desenvolvimento); 2004

[12] Carvalho M. Manual de Reflorestamento. 1st ed. Sagrada Família: Belém; 2006

[13] Pitelli RA. Competição e controle das plantas daninhas em áreas agrícolas. Série Técnica IPEF, Piracicaba. 1987;**4**:1-24

[14] Charles-Dominique P. Inter-relations between furgivorous vertebrates and pioneer plants: Cecropia, birds and bats in French Guyana. In: Estrada A, Fleming ETH, editors. Furgiveres and Seeds Dispersal. Dordrecht; 1986. pp. 119-135

[15] Andrade C, Fontes J, Oliveira T, et al. Reforma de pastagens com alta infestação de capim-navalha (*Paspalum virgatum*). 64th ed. Embrapa Acre-Circular Técnica: Rio Branco; 2012

[16] Albuquerque J. Identificação de plantas mvasoras de cultura da região de Manaus. Acta Amaz. 1980;**10**:47-95

[17] Martins MV, Shimizu GH, Bittrich V. Flora da Reserva Ducke, Estado do Amazonas, Brasil: Hypericaceae. Hoehnea. 2018;**45**:361-371

[18] Flora do Brasil 2020 (em construção). Hypericaceae. Jardim Botânico do Rio de

Janeiro. 2018. http://www.floradobrasil.jbrj.gov.br/reflora/floradobrasil/ FB133 [Accessed: 10 January 2018]

[19] Mourão K, Beltrati C. Morphology and anatomy of developing fruitsand seeds of Vismia guianensis (aubl.) Choisy (Clusiaceae). Revista Brasileira de Biologia. 2001;**61**:147-158

[20] Zenni RD, Ziller SR. An overview of invasive plants in Brazil. Revista Brasileira de Botanica. 2011;**1**:431-446

[21] Snow N, Lau A. Notes on grasses (Poaceae) in Hawaii: 2. Records of the Hawaii Biological Survey. 2008;**2**:46-60

[22] Cruz R, Merayo A, Zuñiga G, et al. *Paspalum virgatum* L. In: Manejo de Malezas Para Países en Desarrollo. Roma; 1996. p. 403

[23] Tropicos.org. Missouri Botanical Garden. 2018. http://www.tropicos.org/Name/13036675. [Accessed: 10 August 2018]

[24] Ahmad H, Sehgal S, Mishra A, et al. *Mimosa pudica* L. (Laajvanti): An overview. Pharmacognosy Reviews. 2012;**6**:115-124

[25] Azmi L, Singh MK, Akhtar AK. Pharmacological and biological overview on Mimosa pudica Linn. International Journal of Pharmaceutical and Life Sciences. 2011;**2**:1226-1234

[26] Ripardo FH, Pacheco L, Andrade E, et al. Phytotoxic activity of compounds from *Moutabea guianensis* aubl. on Amazonian invasive species. Eclética Química. 2015;**40**:71-76

[27] Musah RA, Lesiak AD, Maron MJ, et al. Mechanosensitivity below ground : Touch-sensitive smell-producing roots in the shy plant. Plant Physiology. 2016;**170**:1075-1089

[28] Simon FW, Hodson CN, Roitberg BD. State dependence, personality, and plants: Light-foraging decisions in Mimosa pudica (L.). Ecology and Evolution. 2016;**17**:6301-6309

[29] Dutra V, Morim M. Mimosa in Lista de Espécies da Flora do Brasil. Jardim Botânico do Rio de Janeiro. 2018. http://floradobrasil.jbrj.gov.br/jabot/floradobrasil/FB83449. [Accessed: 10 January 2018]

[30] Baraúna AC, Rouws LFM, Simoes-araujo JL, et al. *Rhizobium altiplani* sp. nov., isolated from effective nodules on Mimosa pudica growing in untypically alkaline soil in Central Brazil. International Journal of Systematic and Evolutionary Microbiology. 2018;**66**:4118-4124

[31] Dourado D, Conceição A, Santos-Silva J. *O gênero Mimosa* L. (Leguminosae: Mimosoideae) na APA Serra Branca/Raso da Catarina, Bahia, Brasil. Biota Neotropica. 2013;**13**:225-240

[32] Volkov AG, Foster JC, Ashby TA, et al. Mimosa pudica: Electrical and mechanical stimulation of plant movements. Plant, Cell and Environment. 2010;**33**:163-173

[33] Reed-Guy S, Gehris C, Shi M, et al. Sensitive plant (*Mimosa pudica*) hiding time depends on individual and state. PeerJ. 2017;**5**(e3598):1-16

[34] Cahill JF, Bao T, Maloney M, et al. Mechanical leaf damage causes localized, but not systemic, changes in leaf movement behavior of the sensitive plant, *Mimosa pudica* (Fabaceae) L. Botany. 2013;**91**:43-47

[35] Mauldin E, Peters-Kennedy J. Integumentary system. In: Pathology of Domestic Animals. Philadelphia: WB Saunders Ltd.; 2016. pp. 509-736

[36] Carvalho A, Carvalho N, Vieira G, et al. Intoxicação espontânea por Senna obtusifolia em bovinos no Pantanal Sul-Mato-Grossense. Pesquisa Veterinaria Brasileira. 2014;**34**:147-152

[37] Doughari J, El-Mahmood A, Tyoyina I. Antimicrobial activity of leaf extracts of *Senna obtusifolia* (L). African Journal of Pharmacy and Pharmacology. 2008;**2**:7-13

[38] Tungate KD, Susko DJ, Rufty TW, et al. Reproduction and offspring competitiveness of Senna obtusifolia are influenced by nutrient availability. The New Phytologist. 2002;**154**:661-669

[39] Takano HK, Constantin J, Braga G, et al. Dry season and soil texture affect the chemical control of Senna obtusifolia in sugarcane. Revista Brasileira de Herbicidas. 2015;**403**:181-193

[40] Dunlop EA, Wilson JC, Mackey AP. The potential geographic distribution of the invasive weed Senna obtusifolia in Australia. Weed Research. 2006;**46**:404-413

[41] Topanotti L, Pereira P, Bechara F. Germinação de sementes de *Senna obtusifolia* (L.) H. S. Irwin & Barneby (Fabaceae) visando a restauração de áreas degradadas. Publicatio UEPG: Ciências Biológicas e da Saúde. 2014;**20**:125-129

[42] Lynn Walker H, Tilley A. Evaluation of an isolate of *Myrothecium verrucaria* from sicklepod (*Senna obtusffolia*) as a potential mycoherbicide agent. Biological Control. 1997;**112**:104-112

[43] Govaerts S, David S. No Title. World checklist of Cyperaceae, Royal Botanic Gardens, Kew [Internet]. 2007. http://www.kew.org/wcsp/monocots/Jan [Accessed: 6 October 2018]

[44] Jakelaitis A, Ferreira L, Silva A, et al. Effects of management systems on purple nutsedge populations (*Cyperus rotundus*). Planta Daninha. 2003;**21**:89-95

[45] Panozzo L, Agostinetto D, Galon L, et al. Cyperus esculentus handling methods in irrigated rice. Planta Daninha. 2009;**27**:165-174

[46] Boulos L, El-Hadidi M. The Weed Flora of Egypt. 2nd ed. Cairo: American University in Cairo Press; 1984

[47] Seo W, Pae H, Oh G, et al. Inhibitory effects of methanol extract of *Cyperus rotundus* rhizomes on nitric oxide and superoxide productions by murine. Journal of Ethnopharmacology. 2001;**76**:59-64

[48] Gupta M, Palit T, Singh N, et al. Pharmacological studies to isolate the active constituents from *Cyperus rotundus* possessing anti-inflammatory, anti-pyretic and analgesic activities. The Indian Journal of Medical Research. 1971;**59**:76-82

[49] Dang GK, Parekar RR, Kamat SK, et al. Anti inflammatory activity of *Phyllanthus emblica*, *Plumbago zeylanica* and *Cyperus rotundus* in acute models of inflammation. Phytotherapy Research. 2011;**25**:904-908

[50] Peerzada A, Ali H, Naeem M, et al. *Cyperus rotundus* L.: Traditional uses, phytochemistry, and pharmacological activities. Journal of Ethnopharmacology. 2015;**174**:540-560

[51] Yu J, Lei G, Cai L, et al. Chemical composition of *C. rotundus* extract. Journal of Phytochemistry. 2004;**65**:881-889

[52] Holm L, Doll L, Holm E, et al. World Weeds. Natural Histories and Distributions. New York, NY: John Wiley & Sons, Inc.; 1997

[53] Quiroz C, Choler P, Baptist F, et al. Alpine dandelions originated in the native and introduced range differ in their responses to environmental constraints. Ecological Research. 2009;**24**:175-183

[54] Martinez M, Poirrier P, Chamy R, et al. Taraxacum of fi

cinale and related species—An ethnopharmacological review and its potential as a commercial medicinal plant. Journal of Ethnopharmacology. 2015;**169**:244-262

[55] Molina-Montenegro M, Atala C, Gianoli E. Phenotypic plasticity and performance of *Taraxacum officinale* (dandelion) in habitats of contrasting environmental heterogeneity. Biological Invasions. 2010;**12**:2277-2284

[56] Molina-Montenegro M, Peñuelas J, Munné-Bosh S, et al. Higher plasticity in ecophysiological traits enhances the performance and invasion success of *Taraxacum officinale* (dandelion) in alpine environments. Biological Invasions. 2012;**14**:21-33

[57] Molina-Montenegro M, Quiroz C, Torres-Díaz C, et al. Ecophysiological traits suggest local adaptation rather than plasticity in the invasive *Taraxacum officinale* (dandelion) from native and introduced habitat range. Plant Ecology & Diversity. 2011;**4**:36-42

[58] Molina-Montenegro M, Palma-Rojas C, Alcayaga-Olivares Y, et al. Ecophysiological plasticity and local differentiation help explain the invasion success of Taraxacum officinale (dandelion) in South America. Ecography. Ecography (Cop). 2013;**36**:718-730

[59] Cavieres LA, Quiroz CL, Molina-Montenegro MA. Facilitation of the non-native *Taraxacum officinale* by native nurse cushion species in the high Andes of central Chile: Are there differences between nurses? Functional Ecology. 2008;**22**:148-156

[60] Giacomino A, Malandrino M, Colombo ML, et al. Metal content in dandelion (*Taraxacum officinale*) leaves: Influence of vehicular traffic and safety upon consumption as food. Journal of Chemistry. 2016;**2016**:9

[61] Maleci L, Buffa G, Wahsha M, et al. Morphological changes induced by heavy metals in dandelion (*Taraxacum officinale* Web.) growing on mine soils. Journal of Soils and Sediments. 2014;**14**:731-743

[62] Malawska M, Wiłkomirski B. An analysis of soil and plant (*Taraxacum officinale*) contamination with heavy metals and policyclic aromatic hydrocarbons (PAHs) in the area of the railway junction Iława Główna, Poland. Water, Air, and Soil Pollution. 2001;**127**:339-349

[63] Królak E. Accumulation of Zn, Cu, Pb and Cd by Dandelion (*Taraxacum officinale* Web.) in environments with various degrees of metallic contamination. Polish Journal of Environmental Studies. 2003;**12**:713-721

[64] Tu M, Hurd C, Randall JM, et al. Weed Control Methods Handbook: Tools & Techniques for Use in Natural Areas. 2001

[65] Doyle CJ. A review of the use of models of weed control in Integrated Crop Protection. 8809

[66] Slaughter DC, Giles DK, Downey D. Autonomous robotic weed control systems: A review. Computers and Electronics in Agriculture. 2008;**6**:63-78

[67] De Souza CP, Guedes TDA, Fontanetti CS. Evaluation of herbicides action on plant bioindicators by genetic biomarkers: A review. Environmental Monitoring and Assessment. 2016;**188**:1-12

[68] Agostinetto D, Vargas L, Gazziero D, et al. Manejo de plantas daninhas. In: Soja: Do Plantio à Colheita. Viçosa: UFV; 2015. pp. 234-255

[69] Paszko T, Muszyński P, Materska M, et al. Adsorption and degradation of phenoxyalkanoic acid herbicides in soils:

A review. Environmental Toxicology and Chemistry. 2016;**35**:271-286

[70] Pannacci E, Tei F. Effects of mechanical and chemical methods on weed control, weed seed rain and crop yield in maize, sunflower and soyabean. Crop Protection. 2014;**64**:51-59

[71] Embrapa Trigo. Métodos de controle de plantas daninhas [Internet]. 2006. http://www.cnpt.embrapa.br/biblio/do/p_do62_8.htm [Accessed: 10 February 2018]

[72] Weed Science Society of America. WSSA Position Statement on Biological Control of Weeds [Internet]. 2018. http://wssa.net/wssa/weed/biologicalcontrol/ [Accessed: 10 April 2018]

[73] Uludag A, Uremis I, Arslan M. Biological Weed Control. 1st ed. Elsevier: Academic Press. 2018. DOI: 10.1016/B978-0-12-809881-3.00007-3

[74] Souza Filho A, Alves S. Alelopatia em Ecossistema de Pastagem Cultivada. 109th ed. EMBRAPA-CPATU: Belém; 1998

[75] Holt J, Lebaron H. Significance and distribution of herbicide resistance. Weed Technology. 1990;**4**:141-149

[76] Ryan G. Resistance of common groundsel to simazine and atrazine. Weed Science. 1970;**18**:614-616

[77] Lebaron H, Mcfarland J. Herbicide resistance in weeds and crops: An overview and prognosis. In: Managing Resistance to Agrochemicals: From Fundamental Research to Practical Strategies. Washington: American Chemical Society; 1990. pp. 336-352

[78] Souza Filho A, Pereira A, Bayma J. Aleloquimico produzido pela gramínea forrageira Brachiaria humidicola. Planta Daninha. 2005;**23**:25-32

[79] Mourão Júnior M, Souza FA. Differences in allelopathic activity patterns in Leguminosae. Planta Daninha. 2010;**28**:939-951

[80] Souza Filho A, Fonseca M, Arruda M. Substâncias químicas com atividades alelopáticas presentes nas folhas de Parkia pendula (Leguminosae). Planta Daninha. 2005;**23**:565-573

[81] Iqbal Z, Furubayashi A, Fujii Y. Allelopathic effect of leaf debris, leaf aqueous extract and rhizosphere soil of Ophiopogon japonicus Ker-Gawler on the growth of plants. Weed Biology and Management. 2004;**4**:43-48

[82] Rice E. Allelopathy. 1st ed. Orlando: Academic Press; 1984

[83] Miller DA. Allelopathy in forage crop systems. Agronomy Journal. 1996;**88**:854-859

[84] Muller CH. The role of chemical inhibition (allelopathy) in vegetation composition. Bulletin of the Torrey Botanical Club. 2013;**93**:332-351

[85] Szczepahiski AJ. Allelopathy as a mean of biological control of water weeds. Aquatic Botany. 1977;**3**:193-197

[86] Wardle DA. Allelopathic in New Zealand pasture grassland ecosystem. New Zealand Journal of Experimental Agriculture. 1987;**15**:243-255

[87] Reigosa MJ, Pazos-Malvido E. Phytotoxic effects of 21 plant secondary metabolites on *Arabidopsis thaliana* germination and root growth. Journal of Chemical Ecology. 2007;**33**:1456-1466

[88] Inderjit I, Callaway R, Vivanco J. Can plant biochemistry contribute to understanding of invasion ecology? Trends in Plant Science. 2006;**11**:574-579

[89] Rizvi S, Rizvi V. Allelopathy: Basic and Applied Aspects. London: Chapman and Hall; 1992

[90] Rizivi S, Rizivi S, Tahir M, et al. Allelopathic interactions in agroforestry systems. Critical Reviews in Plant Sciences. 1999;**18**:773-796

[91] Shimoji H, Yamasaki H. Inhibitory effects of flavonoids on alternative respiration of plant mitochondria. Biologia Plantarum. 2005;**49**:117-119

[92] Sakihama Y, Cohen MF, Grace SC, et al. Plant phenolic antioxidant and prooxidant activities: Phenolics-induced oxidative damage mediated by metals in plants. Toxicology. 2002;**177**:67-80

[93] Inderjit I. Plant phenolics in allelopathy. Botanical Review. 1996;**62**:186-202

[94] Putnam AR. Allelochemicals from plants as herbicides. Weed Technology. 1988;**2**:510-518

[95] Periotto F, Perez S. MIS. L. Efeito alelopático de Andira humilis Mart. ex Benth na germinação e no crescimento de *Lactuca sativa* L. e *Raphanus sativus* L. Acta Botânica Brasílica. 2004;**18**:425-430

[96] Yokotani-Tomita K, Goto N, Kosemura S, et al. 0 10 100 concentration (ppm). Phytochemistry. 1998;**47**:1-2

[97] Yamada K, Anai T, Hasegawa K. Lepidimoide, an allelopathic substance in the exudates from germinated seeds. Phytochemistry. 1995;**39**:1031-1032

[98] Tefera T. Allelopathic effects of *Parthenium hysterophorus* extracts on seed germination and seedling growth of Eragrostis tef. Journal of Agronomy and Crop Science. 2002;**188**:306-310

[99] Inoue M, Santana D, Pereira M, et al. Extratos aquosos de Xylopia aromatica e Annona crassiflora sobre capim-marandu (*Brachiaria brizantha*) e soja. Scientia Agrária. 2009;**10**:245-250

[100] Kato-Noguchi H. Isolation and identification of an allelopathic substance in Pisum sativum. Phytochemistry. 2003;**62**:1141-1144

[101] Clout MN, Williams PA. Invasives Species Management: A Handbook of Principles and Techniques. New York: Oxiford; 2009

[102] McFadyen RE, Harvey GJ. Distribution and control of rubber vine, Cryptostegia grandiflora, a major weed in northern Queensland. Plant Protection Quarterly. 1990;**5**:152-155

[103] Evans HC, Tomley AJ. Studies on the rust, Maravalia cryptostegiae, a potential biological control agent of rubber-vine weed, Cryptostegia grandiflora (Asclepiadaceae: Periplocoideae), in Australia, III : Host range. Mycopathologia. 1994;**126**:93-108

[104] Taylor CAZM, Hastings A. Finding optimal control strategies for invasive species: A density-structured model for Spartina alterniflora. Journal of Applied Ecology. 2004;**41**:1049-1057

[105] Sobhian R, Littlefield J, Cristofaro M, et al. Biology and host specificity of Spurgia capitigena (Bremi) (Dipt., Cecidomyiidae), for the biological control of Euphorbia esula L. in North America. Journal of Applied Entomology. 2000;**124**:333-338

[106] Winston RL, Schwarzländer M, Hinz HL, et al. Biological Control of Weeds: A World Catalogue of Agents and their Target Weeds. 2014

[107] Souza-Alonso P, Lorenzo P, Rubido-Bará M, et al. Forest Ecology and Management Effectiveness of management strategies in Acacia dealbata Link invasion , native vegetation and soil microbial community responses. Forest Ecology and Management. 2013;**304**:464-472

[108] McDonald PM, Abbott CS, Fiddler GO. Density and development of bracken fern (Pteridium aquilinum) in Forest plantations as affected by manual and chemical application. Native Plants. 2003;**4**:52-60

[109] Hempy-Mayer K, Pyke DM. Defoliation effects on bromus tectorum seed production: Implications for grazing. Rangeland Ecology & Management. 2008;**61**:116-123

Section 2

Impact of Invasive Species on Ecosystem and Its Control

Chapter 3

Impacts of Invasive Plants on Soil Fungi and Implications for Restoration

Brooke Pickett, Mia Maltz and Emma Aronson

Abstract

Biological plant invasions impact the function and biodiversity of ecosystems across the globe by displacing native plant species and altering the physical and chemical soil environment. While much is known about direct competition between invasive and native plants, ecologists have just begun to uncover the less obvious impact of plant invasion: changes to the soil fungal community. Fungi are important to the survival of many plant species and an integral part of a healthy soil system. Arbuscular mycorrhizal fungi are plant mutualistic symbionts that associate with many species and provide necessary services, such as increasing surface area for root water absorption and resistance to pathogens, while ectomycorrhizal fungi play an equally important role and are critical for plant nutrient acquisition in boreal and temperate forests. Invasive plants are altering the soil fungal community in ways that indirectly impact the structure of native plant communities, sometimes for years after the invasive plant has been removed from an area (i.e., legacy effects). These changes make restoration especially difficult in areas from which long-term plant invasions have been eradicated; in some cases these changes can be so severe that even with active management, they take months or decades to reverse.

Keywords: mycorrhizal, fungi, roots, legacy effects, restoration, microbial, invasion

1. Introduction

The global scale of plant invasion means we need to understand it better at all levels in order to prevent further damage. While much research has been conducted about the ecosystem impact of invasive plants, ecologists have recently begun to uncover a less obvious, but important, consequence of plant invasion: changes to the soil fungal community.

Fungi are ubiquitous and the principal decomposers of organic debris in ecosystems all around the world [1]. They are essential to decomposition and nutrient cycling in most intact environments, ranging from unicellular aquatic chytrids to large mushroom fruitbodies with extensive mycelial networks. They acquire their food by exuding enzymes into their environment, breaking apart the bond structures in complex compounds, and subsequently absorbing the dissolved nutrients and molecular components. Some fungi exist as symbionts of plants and animals while others exist as free-living cells. Symbionts can interact with their host as mutualists, parasites, or in a way that does not affect the host (commensalism) [1]. Commensalism, in this context, not only includes symbionts but also free-living microorganisms performing nutrient transformations critical to plant growth, such as nitrification and denitrification [2].

Fungal mutualists interact with plants through mycorrhizal symbiosis, a symbiotic association between fungal hyphae and the roots of a vascular plant that can be characterized as either arbuscular mycorrhizal, ectomycorrhizal, or ericoid [3]. These mycorrhizal fungi grow in the rhizosphere of the plant and can be either intracellular (arbuscular mycorrhizal fungi; AMF) or extracellular (ectomycorrhizal; ECM). Plants and their symbionts communicate through molecular and genetic feedback during which fungi provide growth-limiting nutrients, such as nitrate and phosphate [3], and even facilitate plant-to-plant exchange of nutrients and carbohydrates [4]. These plant-fungal associations are extremely important to the survival of a majority (~90%) of all plant species [52].

When invasive plants are introduced to a healthy ecosystem, they can disrupt fungal mutualistic associations with native plants. Moreover, plant invasions may even prevent these mutualistic associations from occurring by altering soil nutrient dynamics, changing soil food webs, or introducing plant pathogens [2, 3]. Although not always negatively impacting mutualisms with native plants, these changes brought on by plant invasion can last for years [2, 4–10] after the invasive plant has been removed and are termed "legacy effects" [13]. These legacy effects are normally defined as the abiotic and biotic impact of a species that persist long after the invasive species has been eradicated or extirpated from an area [8].

Studies focused on understanding the legacy effects of invasive plant growth on native plants can have either similar or conflicting results, largely dependent upon the native and invasive species studied [14]. As a result, many suggestions for improving soils after invasive species removal have been anecdotal, and are context-dependent.

In this chapter, we will discuss how invasive plants may change the abundance or diversity of three important fungal symbionts (arbuscular mycorrhizal fungi, ectomycorrhizal fungi, and fungal pathogens), as well as the implications these changes may have for ecosystem health. We will finish off the chapter by discussing restoration efforts designed to ameliorate fungal legacy effects of invasive plants.

2. Biotic impact of invasive plants

Plant-soil interactions can be abiotic or biotic, meaning that plant composition can alter the chemical composition of the soil or the microbial composition of the soil and vice versa. Not until 1985, however, did papers linking the words plant and soil begin to appear in the BIOSIS database. Since then, papers about plant-soil interactions have appeared at a rate of 3500 per year [2]. So while the field is relatively new, it is growing quickly and becoming more diverse.

Early investigations of plant-soil feedbacks focused on physical properties of the soil, such as texture, water content, and temperature. Researchers then began investigating the chemical and biogeochemical components of plant-soil feedbacks, such as the pH, carbon, and nitrogen content of soils [2]. Currently, there is more focus on the role of microbes in regulating and responding to plants and the larger environment. This increased focus on microbes is due to their critical importance to the ecology of all macro-organisms: they are major decomposers in all ecosystems, important to the survival of most plant species, and an integral part of both carbon and nitrogen cycles.

Many studies have demonstrated shifts in microbial communities due to invasive plant growth [4, 12–17]. However, the phenomenon of *fungal* shifts in response to invasive plants is less understood, and potentially has many implications for maintaining biodiversity and function of invaded ecosystems. Throughout this section we will explore the ways in which invasive plants alter the fungal community (**Figure 1**).

2.1 Arbuscular and ectomycorrhizal fungi

Fungal hyphae, or collectively the mycelium or mycelial network, are filamentous strands of fungal cells which compose the main body of the fungus, and the fungal vegetative structure that is often branching and filamentous [20]. In soils, fungal hyphae grow throughout the soil matrix, with the direction of apical growth (from the apex to the hyphal tip) often dependent on an environmental stimulus. These hyphae exhibit a variety of morphological structures and functional modifications. Arbuscular mycorrhizal fungi (AMF) form arbuscules, small branching structures within cortical root cells, which are the sites of the bi-directional exchange of carbon and nutrients, such as phosphorus, between the plant and fungi [20]. Ectomycorrhizal fungi similarly exchange nutrients with plants, but they form a dense hyphal sheath that surrounds the root surface, rather than penetrating the root cells [21] (**Figure 2**). This mutualism provides a fungus with carbohydrates and the plant with an increased surface area for water and mineral absorption.

Arbuscular mycorrhizal fungi are obligate plant symbionts. These AMF are arguably the most common plant mutualistic symbionts, consisting of at least 145 groups [22]. They associate with most plant species and are especially important for the uptake of phosphorus [16, 18], an integral nutrient for plant growth. Over the years, researchers have discovered that AMF not only increases plant access to phosphorus, but they also provide resistance to pathogens [18, 19], stabilize soil

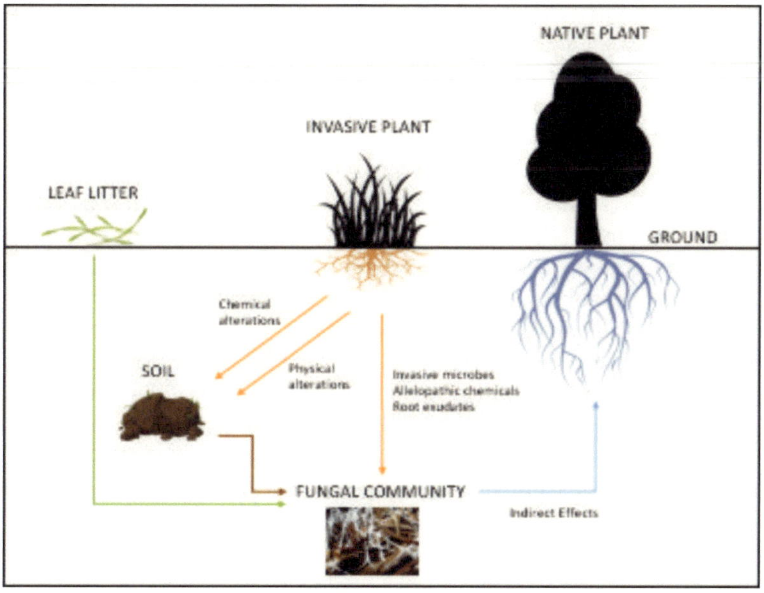

Figure 1.
Diagram showing the biotic impacts of invasive plants. Orange arrows: *the invasive plant alters the chemical and physical soil components, which has an indirect effect on the fungal community composition. The invasive plant directly affects the fungal community through introduction of invasive microbes, allelopathic chemicals, and root exudates.* Green arrow: *leaf litter can alter the fungal community composition if the invasive plant leaf litter has a different quality (C:N) than that of the native plant leaf litter.* Blue arrow: *all of these alterations to the soil fungal community have indirect effects on the growth of native plants.*

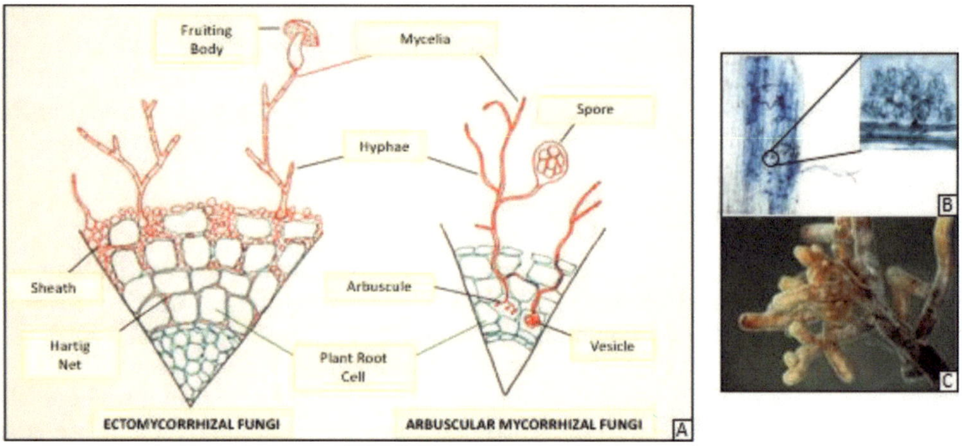

Figure 2.
(A) Diagram depicting the similarities and differences between ectomycorrhizal and arbuscular mycorrhizal fungi. (B) An arbuscule inside of a plant root. (C) Ectomycorrhizal fungal hyphae growing on a plant root.

aggregates [23], alter plant communities [22], and even ameliorate the allelopathic effect of some invasive plants [24]. Most AMF are generalists, meaning they associate with many plant taxa, while others are specialists, and associate with only one or merely a few plant taxa.

The widespread distribution and low host-specificity of most AMF suggests that when plants invade a healthy soil system, they can readily form associations with AMF. Since most invasive plants can probably form arbuscular mycorrhizas [25], it is not surprising to find that numerous opportunistic invasive plants also associate with AMF to their own advantage [26]. When associating with fast-growing, small-spored fungal taxa, such as *Glomus*, which can colonize via mycelia fragments, an invasive plant may be even more likely to thrive [27]. These associations with generalist AMF may allow invasive plants to outcompete and displace native plants which are either non-mycorrhizal (such as *Brassica* spp.), weakly mycorrhizal, or do not form associations with generalist AMF, in contrast to the generalist invader. One recent example of such an invader is *Vincetoxicum rossicum*, a forb that displaces native plants and was found to associate with four different AMF subgroups (*Glomus intraradices*, *G. caledonium*, *G. fasciculatum*, and *G. mosseae*), which are highly infective and remarkably efficient at phosphorus uptake. These same subgroups, however, were absent from the rhizosphere of each native plant growing within the invasive plant patches [27]. This finding suggests that the invasive plant's ability to associate with fungal generalists allows it to thrive and may improve its ability to displace native plants.

Some invasive plants have the ability to degrade local mycorrhizal fungi, a finding termed the "Mycorrhizal Degradation Hypothesis" [28] (**Figure 3**). Degradation of local AMF can change the soil in ways that hinder native plants and help invasive plants. Examples of this include *Alliaria petiolata*, a non-mycorrhizal plant, which has been known to produce glucosinolates which are potentially toxic to AMF, and *Myrica faya*, a plant which carries nitrogen fixing microbes, from the genus *Frankia* along with it to the invaded range [29].

While researchers have only just begun exploring the impact of invasive plant species on AMF abundance [25–27], it is evident that invasive plants can have the potential to either increase [14, 28, 29] or decrease [29, 30] the abundance and diversity of AMF. Increased AMF abundance with invasion may happen when the native intact plant community naturally associates with fewer AMF taxa than the invading mycotrophic (mycorrhizal) plants [4, 14, 24, 31]. In fact, if the invader is mycotrophic, a monoculture of the invasive plant may still harbor a more species-rich AMF

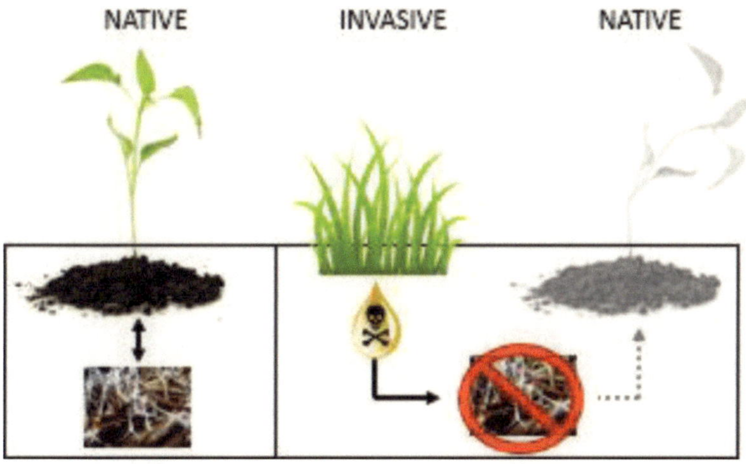

Figure 3.
Diagram illustrating the mycorrhizal degradation hypothesis. In the left panel, we see a healthy native plant in a mutualistic relationship with AMF. In the right panel, the invasive plant is producing a chemical exudate that eliminates beneficial fungi, thereby preventing fungal association with the native plant and eventually native plant death.

community than a diverse community of native plant species [17]. This increased abundance of AMF by the invader may actually feedback to increase invasion [36]. However, if the invading plant is non-mycorrhizal, then AMF abundance and diversity will decrease relative to pre-invasion soil [37]. A recent comprehensive field study [32] compared AMF abundance in soils invaded by one non-mycorrhizal and two mycorrhizal plant species. All three invaders reduced AMF abundance and richness, but the non-mycorrhizal plant reduced AMF abundance and richness to a greater extent. However, this pattern is not always so evident: if an invader is mycotrophic, but not a good host for AMF, then it may actually decrease the AMF abundance [33, 34]. Certain invasive plants may associate with particular groups of AMF, which may be different than those hosted by local native plants [39]. In these cases, invasion could subsequently bolster the abundance of some AMF groups, while decreasing the diversity or abundance of others.

In the presence of invasive plants, some studies show shifts in either AMF diversity [11] or from fungal specialists to generalists [3, 35], as well as differences in the prevalence of fungal versus bacterial utilization of leaf litter [11]. However, the identity and functional group status of both the native and invasive plant may dictate their effects on AM fungal symbionts. For instance, a recent meta-analysis [31] reported that invasions may not necessarily cause a shift in AMF associations, unless the native and invasive plant are in different functional groups. If an invader decreases AMF abundance or only increases the abundance of the particular AM fungal associate, then this could negatively impact native plant communities which are dependent on AMF for survival [31]. Changes to soil AMF abundance and diversity may not be short-lived; in fact, they could last long after the invader is gone [41]. Such biotic legacy effects can occur when plant-soil interactions are altered by invasive plants for long periods of time.

The timing of AMF response to invasion is still largely a mystery [42]. A recent meta-analysis [31] reported that AMF colonization of native plants may decrease due to legacy effects of invasive plants. However, it is unclear how quickly these legacy effects occur or attenuate after an invasive plant is removed, as well as how soon the community may return to the structure and functioning of the previously native state [42]. In certain instances, after an invasive is removed, any changes in AMF abundance and diversity are fleeting, because differences in abundance and richness return rapidly with the return of the native vegetation type [14, 38]. In contrast, in other studies [37] even a highly mycorrhizal invasive plant may not rapidly alter the AMF community, even after 29 weeks. Another study shows some recovery of AMF communities 6 years after the removal of an invasive known to decrease AMF abundance [13]. In some cases invasion can lead to the development of a novel AMF community over decadal time scales [30, 37, 39]. Overall, the recovery of the AMF community could take a long time. Furthermore, shifts in AMF may be dependent on an invasive plant's functional traits [39–42], which may ultimately be the best predictor for the extent of AMF response to invasion, and subsequent recovery.

Although AMF and other groups of mycorrhizal fungi, such as the ectomycorrhizal fungi (ECM or EM fungi), are phylogenetically distant [45] and functionally distinct, they both play key roles in ecosystem functioning. In boreal and temperate forests, ECM are facultative symbionts that play an important role in plant nutrient acquisition [46]. In fact, ECM take up about 80% of all plant nitrogen in boreal forests [47]. However, the impact of invasive plants on the soil composition of ECM has not been well studied [48]. The few papers that do tackle this issue have either found inhibition of ECM in the presence of a non-mycorrhizal invader [43–45] or suggest that an invasive plant may elicit an allelopathic effect on EM fungi [50]. Similarly as with AMF, ECM associations with native plants can be inhibited by invasive plant presence [49].

Invasive plants also introduce invasive AMF and ECM into the invaded range. Very little is known about the invasion process of AMF, but we do know that AMF propagules can be transferred long distances by wind, water, and agriculture [51]. When AMF is introduced to a new area, it spreads very slowly from the point of introduction, but can persist for up to a several years in the soil without a host [51]. Introduction could be problematic if the AMF are generalists and associate with invasive plants; in these cases it may not have an overall negative impact on an ecosystem that already harbors AMF. Normally ECM are beneficial to plants, but they have been shown to cause damage to invaded ecosystems by competing with native fungi, facilitating in the co-invasion of trees [52], and changing the soil foodwebs [53]. It is still unknown, however, what effect these invaders have on native host physiology and native fungal communities.

2.2 Fungal pathogens

Soil pathogens contribute to the spatial and temporal patterns of natural systems through negative plant-soil feedbacks [54] and may influence plant diversity by suppressing dominant plants [55]. Certain pathogens target either a group of related plant species or only one host plant genus.

Increase in global trade and the subsequent movement of plants has increased the number of introduced plant species and the pathogens they carry [56]. Some invasive pathogens have been introduced intentionally as biological controls [57], but most may be introduced inadvertently over trade routes. The fact that fungi are small and inconspicuous may be a major factor in their success as invaders and may be why these pathogens can spread faster than the host plants that carry them.

Pathogens brought over by invaders have been shown to decimate native plant populations. The lack of host resistance to invasive pathogens has caused severe environmental and agricultural damage in invaded areas [1]. Some examples of pathogenic microbes, often studied by plant pathologists and mycologists, are *Phytophthora cinnamomi* which infects Eucalyptus trees in Australia [58], *Armillaria luteobubalina* which has killed off 38% of plants in coastal ecosystems [59], *Phytophthora ramorum* which has infected more than 70 plant species in California and causes sudden oak death [60], and *Phytophthora kernoviae* which is the latest of many *Phytophthoras* recently found in the UK [56] (**Figure 4**).

Figure 4.
(A) Phytophthora cinnamomi*, (B)* Phytophthora kernoviae*, (C)* Armillaria luteobubalina*, and (D)* Phytophthora ramorum.

Most studies concerning the spread of invasive fungal pathogens focus primarily on agricultural rather than natural systems. In a majority of papers, the invading pathogen that causes a devastating agricultural epidemic are those that coevolved with crop plants and were somehow reunited with their host [1]. The most well-known example of this is the Irish potato famine caused by *Phytophthora infestans*. In contrast, in natural systems, most harmful invasive pathogens did not coevolve with the plants they infect so the host plants have never been exposed to the pathogen before [1].

The main body of research that does focus on invasive pathogens in natural systems primarily deals with invasive forest pathogens. North American forests are continually threatened by invasive pathogens and several species of trees have already been essentially eliminated by them. The best example of this is the chestnut blight which killed off most of the mature native chestnut trees in the northeast US in only 30 years [61], replacing them with a variety of other hardwood species. Pests and pathogens may be even more harmful to these hardwood forests than the invasive plants that carried them there [62], with over 20 invasive pathogens infecting forests in the US and Canada [63].

Invasive pathogens in forest ecosystems are currently in the process of removing several foundation tree species that control productivity, water levels, forest structure, and microclimate [64]. When an entire species of tree is wiped out or an entire life stage or size class of tree is eliminated, the forest ecosystem can change dramatically. The loss of these species can negatively impact nutrient fluxes, water movement, biodiversity, and food webs [65]. The indirect effects of these species losses are difficult to calculate and could extend for multiple forest generations. If a relatively minor tree species is lost, the impact of the invasive pathogen may actually be small, but if a keystone species is lost there could be long-lasting cascading effects [64]. It is important to note that not all introduced pathogens are harmful to these forests [66], but more research is needed to identify those that are harmful before they spread.

The damage wrought by invasive pathogens is clearly wide-spread. Approximately 65–85% of plant pathogens are considered invasive [63]. Thus, there is a critical need for invasive pathogen ecology to elucidate the extent to which invasive pathogens harm natural systems. Based on our current understanding, it is unclear whether or not invasive fungal pathogens persist in the soil for years after invasive plants have been removed or whether these pathogens interact with other microbes in the soil to the detriment of native species. In other words, more research is needed to bridge the gap between plant pathology and ecology to better understand the impact of invasive pathogens in natural systems [1].

Introduction of non-native pathogens is one way invasive plants influence soil pathogen composition. However, they have also been shown to influence the abundance and diversity of native fungal pathogens in invaded sites, in ways that are often either beneficial [67] or detrimental to their growth. Alkaloids produced by these pathogens can inhibit generalist pathogenic fungi, which inadvertently stimulates the growth of host-specific pathogens [68]. This accumulation of pathogens specific to the invasive plant may actually allow native plants to thrive [69]. Some studies have shown that certain invasive grasses may produce chemicals which are said to have an inhibitory effect on competitors and may deter herbivory or either repel pathogens [25, 66]. In contrast, some invasive plants may release chemicals known to attract pathogens (Accumulation of Local Pathogens Hypothesis) [71], which could act as a "pathogen reservoir," leading to reduced competition by local plants [72].

3. Restoration efforts to reverse biotic changes

Plant communities are dependent upon soil microbial communities; therefore, native plant restorations may ultimately not be successful unless the microbial and plant communities are simultaneously restored. The idea of using microbes, either a component of the native-plant associated microbial community or an entire whole soil inoculum isolated from an intact ecosystem, as a biological control against the spread of invasive species has gained popularity in recent years.

Restoration ecologists are now applying AMF cultures [37], whole native soil, or biological crust to their restorations in hopes of augmenting native plant establishment (**Figure 5**). Addition of native soil to restoration sites has been found in some studies to decrease invasive plant cover and increase the native plant cover [73]. It

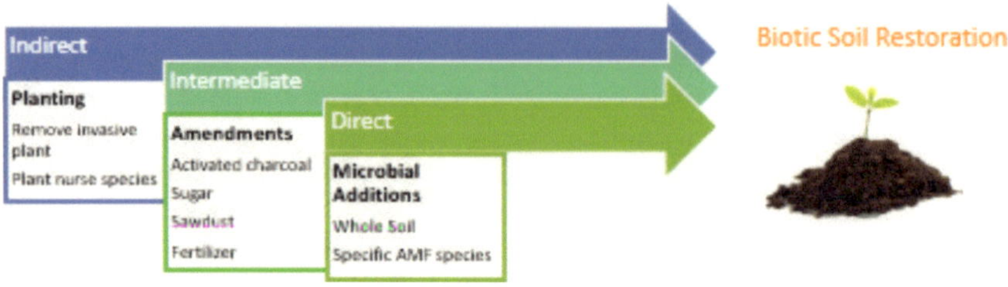

Figure 5.
Diagram showing the indirect, intermediate, and direct methods of biotic soil restoration. Indirect methods include removing invasive plants and planting nurse species. Intermediate methods include soil amendments such as activated charcoal, sugar, sawdust, and fertilizer. Direct methods include the addition of whole soil or specific AMF species to the soil.

is important to compare the methods of these types of studies to understand what inoculation method is most successful for combatting a particular invader [6, 69, 70]. Some studies, for example, remove the invasive plant before applying soil inoculum to restore soil fungal abundance [10] or combine fertilizer with the inoculum. Microbial soil inoculations have been found to actually inhibit the allelochemical effects of an invasive plant on a native plant species [75].

Other possible means of managing invasive plants at the microbial level include the addition of sugar, sawdust, or activated charcoal to soils. Sugar and sawdust can increase microbial growth and store excess soil nitrogen from invasive plants in the microbial biomass [76] (**Figure 5**). This method has been successful for some invasive sites, but not all [9]. Soil additions of activated carbon are believed to bind invasive plant allelochemicals and remove them from the soil solution [77]. Because allelochemicals are short-lived, this technique is most useful if the invasive plant is still present in a site [9]. Studies have shown that native plant growth increases with the addition of activated charcoal under invasion by spotted knapweed (*Centaurea maculosa*), diffuse knapweed (*Centaurea diffusa*), and cheatgrass (*Bromus tectorum*) [9]. Activated carbon can have numerous other effects on the soil (binding organic substrates, changing soil nitrogen concentration, and changing the carbon-to-nitrogen ratio), [9] so further research is needed to decouple these effects with the aforementioned binding of allelochemicals.

3.1 Restoration of arbuscular mycorrhizal fungi and ectomycorrhizal fungi

Invasion by non-mycorrhizal plants can sometimes reduce the abundance of AMF in the soil, negatively impacting native plants that are dependent on AMF for survival. A decrease in AMF abundance can encourage further invasion by non-mycorrhizal plants, thus maintaining invasive plant dominance and inhibiting native plant growth [78]. This is of special concern considering many other studies have found invasive species that are less dependent than native plants on AMF [12, 21, 73–75]. In situations where the invader is known to be non-mycorrhizal, restoration strategies that increase the soil AMF abundance could be especially effective combined with native plant seeding and planting AM host plants. AMF addition to soil has been useful in some restorations efforts [79], but not all [80]. In some cases, when an invaded site has sufficient AM propagule pressure, adding additional AMF may not have any effect on AM abundance or native plant performance [77]. Furthermore, a singular increase in AMF abundance may not be sufficient for restoring native plant diversity, but rather an increase in AMF diversity along with augmenting specialist AMF propagule pressure may improve restoration outcomes [22].

Although some land managers consider the co-invasion of ectomycorrhizal fungi to be a threat to native communities, as of yet there have been minimal evidence-based management strategies documented by practitioners [53]. Removal of plants that associate with ectomycorrhizal fungi, chemical sprays, and sporocarp removal have been performed, but the success of these strategies is debatable. Picking mushroom caps has been shown to have little impact on invasive fungal populations [81], but this may be because studies have focused mainly on fungi that are neither short-lived nor reproduce sexually [53]. Fungicide is another option, but it may also damage native fungi, thereby doing more harm than good to native plants.

3.2 Fungal pathogens and implications for restoration

Very few papers recommend restoration strategies for mitigating the effects of invasive pathogens [53] and even less recommend strategies for preventing the accumulation of pathogens by invasive plants. Most restoration strategies for combatting pathogens are primarily focused on agriculture, not natural systems, and those that do cover natural systems focus primarily on hardwood forests. Restoration of chestnut trees has been extensively studied in the wake of the aforementioned chestnut blight fungus.

Recommended strategies include planting blight resistant trees [82], creating strains of blight fungus that are less virulent [83], and crossbreeding trees [84], such as naturally resistant Asian chestnut trees and American chestnuts. Although many papers focus on gene manipulation as a restoration strategy, others suggest more large-scale strategies such as maintaining tree stand structure, maintaining healthy and resistant tree species, and timber extraction [85]. Some researchers recommend inoculating specific ectomycorrhizal fungi to boost the vigor of infected trees [86]. Blight fungus and hardwood tree infections could potentially be used to guide further research about ecological restoration in other natural systems ravaged by invasive pathogens.

The most successful strategy for combatting invasive fungi is to prevent them from being introduced in the first place. This could involve either banning plants that associate with known invasive fungi or by preventing nurseries from inoculating their plants with invasive fungi [52].

4. Conclusion

Arbuscular and ectomycorrhizal fungi play important roles in the nutrient acquisition and maintenance of biodiversity. Evidence concerning the impact of invasive plants on these fungal groups has been mixed, with AMF occasionally illustrating an increase in abundance [14, 28, 29], a decrease in abundance [39], or a shift from specialist to generalist AM taxa [3, 35]. These conflicting results underscores the importance of future research on the response of AMF to invasion and invasive plant management, with an emphasis on the role of factors driving their response, such as invasive plant functional group [35].

Although little is known about ECM, highlighting a need for future investigation, the evidence suggests that both non-mycorrhizal invasive plants [43–45] and allelopathic invasive plants [50] may inhibit their EM fungal growth, which may interfere with plant nutrient acquisition in both boreal and temperate forests. The introduction of harmful invasive ECM which facilitate in the co-invasion of trees may further disrupt forest symbioses [53].

Invasive plant encroachment into ecosystems have unintended consequences for microbial pathogens, such as influencing the abundance and diversity of native fungal pathogens in ways that benefit their growth or harm native plants. Indeed, invasive plants alter soil fungal composition [63, 67, 82]. The spread of invasive pathogens by invasive species has been widely covered in agricultural research. Future research should focus on invasive pathogens that are being transported by an invasive host plant to natural systems other than hardwood forests. A majority of studies focused on natural systems, emphasize primarily hardwood forests and the loss of foundation tree species. The loss of these foundation species has impacted nutrient fluxes, water movement, biodiversity, and food webs of infected forests [65]. These sorts of large-scale changes could have cascading effects that last for many generations.

Conflicting results and a lack of microbial data has led to case-dependent, anecdotal restoration recommendations. The results of inoculation experiments are very encouraging for improving restoration efforts. However, it may be particularly useful in the future to evaluate exactly how the microbial composition changes for each invasive plant, especially at the species level or for plant functional groups.

While more restoration ecologists are making decisions based on important microbial-plant mutualisms, much more information is needed concerning the long-term impact of invasion on fungi, especially mycorrhizal fungi and fungal pathogens.

Acknowledgements

We are supported by the USDA NIFA AFRI grant CA-R-PPA-5101-CG, USDA NIFA HATCH grant CA-R-PPA-5093-H, and NSF ICER-1541047 and by the University of California President's Research Catalyst Award number CA-16-376437. Also, BP was supported by the National Science Foundation Graduate Fellowship DGE-1326120 and the UCR Center for Conservation Biology Shipley Skinner fund. We thank K. Arogyaswamy, S. Saroa, S. Houssainy, D. Pickett and J. Valliere for intellectual feedback and insightful comments on previous drafts. No conflicts of interest have been declared.

Author details

Brooke Pickett*, Mia Maltz and Emma Aronson
University of California, Riverside, California, United States of America

*Address all correspondence to: brookepic22@gmail.com

References

[1] Desprez-Loustau ML, Robin C, Buee M, Courtecuisse R, Garbaye J, Suffert F, et al. The fungal dimension of biological invasions. Trends in Ecology & Evolution. 2007;**22**(9):472-480

[2] Ehrenfeld JG, Ravit B, Elgersma K. Feedback in the plant-soil system. Annual Review of Environment and Resources. 2005;**30**(1):75-115

[3] Allen MF. The Ecology of Mycorrhizae. Cambridge: Cambridge University Press; 1991

[4] Simard SW, Perry DA, Jones MD, Myrold DD, Durall DM, Molina R. Net transfer of carbon between ectomycorrhizal tree species in the field. Nature. 1997;**388**(6642):579-582

[5] Jordan NR, Larson DL, Huerd SC. Soil modification by invasive plants: Effects on native and invasive species of mixed-grass prairies. Biological Invasions. 2008;**10**(2):177-190

[6] Belnap J, Phillips S, Sherrod S, Moldenke A. Soil biota can change after exotic plant invasion: Does this affect ecosystem processes? Ecology. 2005;**86**(11):3007-3017

[7] Pringle A, Bever JD, Gardes M, Parrent JL, Rillig MC, Klironomos JN. Mycorrhizal symbioses and plant invasions. Annual Review of Ecology, Evolution, and Systematics. 2009;**40**(1):699-715

[8] Cuddington K. Legacy effects: The persistent impact of ecological interactions. Biological Theory. 2012;**6**(3):203-210

[9] Eviner VT, Hoskinson SA, Hawkes CV, Eviner BVT. Ecosystem impacts of exotic plants can feed back to increase invasion in western US rangelands. Rangelands. 2010;**32**(1):21-31

[10] Hamman ST, Hawkes CV. Biogeochemical and microbial legacies of non-native grasses can affect restoration success. Restoration Ecology. 2013;**21**(1):58-66

[11] Hawkes CV, Belnap J, D'Antonio C, Firestone MK. Arbuscular mycorrhizal assemblages in native plant roots change in the presence of invasive exotic grasses. Plant and Soil. 2006;**281**(1-2):369-380

[12] Elgersma KJ, Ehrenfeld JG. Legacy effects overwhelm the short-term effects of exotic plant invasion and restoration on soil microbial community structure, enzyme activities, and nitrogen cycling. Oecologia. 2011:**167**;733-745

[13] Kulmatiski A, Beard KH. Long-term plant growth legacies overwhelm short-term plant growth effects on soil microbial community structure. Soil Biology and Biochemistry. 2011;**43**(4):823-830

[14] Bozzolo FH, Lipson DA. Differential responses of native and exotic coastal sage scrub plant species to N additions and the soil microbial community. Plant and Soil. 2013;**371**(1-2):37-51

[15] Callaway RM, Thelen GC, Barth S, Ramsey PW, Gannon JE. Soil fungi alter interactions between the invader centaurea maculosa and north american natives. Ecology. 2004;**85**(4):1062-1071

[16] Lankau EW, Lankau RA. Plant species capacity to drive soil fungal communities contributes to differential impacts of plant-soil legacies. Ecology. 2014;**95**:3221-3228

[17] Lekberg Y, Gibbons SM, Rosendahl S, Ramsey PW. Severe plant invasions can increase mycorrhizal fungal abundance and diversity. The ISME Journal. 2013;**7**(7):1424-1433

[18] Batten KM, Scow KM, Davies KF, Harrison SP. Two invasive plants alter soil microbial community composition in serpentine grasslands. Biological Invasions. 2006;**8**(2):217-230

[19] Klironomos J. Feedback with soil biota contributes to plant rarity and invasiveness in communities. Nature. 2002;**417**:67-70

[20] Parniske M. Arbuscular mycorrhiza: The mother of plant root endosymbioses. Nature Reviews. Microbiology. 2008;**6**:763-775

[21] Hock B. Fungal Associations. 9th ed. Berlin: Springer; 2012

[22] Bever JD, Schultz PA, Pringle A, Morton JB. Arbuscular mycorrhizal fungi: More diverse than meets the eye, and the ecological tale of why. Bioscience. 2001;**51**(11):923-931

[23] Miller RM, Jastrow JD. Mycorrhizal fungi influence soil structure. Arbuscular Mycorrhizas: Physiology and Function. 2000;**2000**:3-18

[24] Barto K, Friese C, Cipollini D. Arbuscular mycorrhizal fungi protect a native plant from allelopathic effects of an invader. Journal of Chemical Ecology. 2010;**36**:351-360

[25] Richardson DM, Allsopp N, Antonio CMD, Milton SJ, Rejma M. Plant invasions—The role of mutualisms. Biological Reviews. 2000;**75**:65-93

[26] Smith LL, DiTommaso A, Lehmann J, Greipsson S. Effects of arbuscular mycorrhizal fungi on the exotic invasive vine pale swallow-wort (*Vincetoxicum rossicum*). Invasive Plant Science and Management. 2008;**1**:142-152

[27] Bongard C. Fungal colonization of the invasive vine *Vincetoxicum rossicum* and native plants. Plant Ecology, Evolution. 2013;**146**:45-52

[28] Vogelsang KM, Bever JD, Griswold M, Schultz PA. The use of mycorrhizal fungi in erosion control applications. Final Report. Caltrans Contract No. 65A0070, no. 65; 2004

[29] van der Putten WH, Klironomos JN, Wardle DA. Microbial ecology of biological invasions. The ISME Journal. 2007;**1**(1):28-37

[30] Zhang S, Jin Y, Tang J, Chen X. The invasive plant *Solidago canadensis L.* suppresses local soil pathogens through allelopathy. Applied Soil Ecology. 2009;**41**:215-222

[31] Tanner RA, Gange AC. The impact of two non-native plant species on native flora performance: Potential implications for habitat restoration. Plant Ecology. 2013;**214**:423-432

[32] Greipsson S, Ditommaso A. Invasive non-native plants alter the occurrence of arbuscular mycorrhizal fungi and benefit from this association. Ecological Restoration. 2006;**24**(4):236-241

[33] Shah MA, Reshi ZA, Rasool N. Plant invasions induce a shift in Glomalean spore diversity. Tropical Ecology. 2010;**51**:317-323

[34] Mummey DL, Rillig MC. The invasive plant species *Centaurea maculosa* alters arbuscular mycorrhizal fungal communities in the field. Plant and Soil. 2006;**288**(1-2):81-90

[35] Bunn RA, Ramsey PW, Lekberg Y. Do native and invasive plants differ in their interactions with arbuscular mycorrhizal fungi? A meta-analysis. Journal of Ecology. 2015;**103**(6):1547-1556

[36] Paudel S, Baer SG, Battaglia LL. Arbuscular mycorrhizal fungi (AMF) and success of *Triadica sebifera* invasion in coastal transition ecosystems along the northern Gulf of Mexico. Plant and Soil. 2014;**378**:337-349

[37] Zubek S, Majewska ML, Błaszkowski J, Stefanowicz AM, Nobis M, Kapusta P. Invasive plants affect arbuscular mycorrhizal fungi abundance and species richness as well as the performance of native plants grown in invaded soils. Biology and Fertility of Soils. 2016;**52**(6):879-893

[38] Endresz G, Somodi I, Kalapos T. Arbuscular mycorrhizal colonisation of roots of grass species differing in invasiveness. Community Ecology. 2013;**14**:67-76

[39] Busby RR. Cheatgrass (*Bromus tectorum L.*) interactions with arbuscular mycorrhizal fungi in the North American steppe: Prevalence and diversity a associations, and divergence from native vegetation. Dissertation Abstracts International. 2011;**72**(11):1-136

[40] Belnap J, Phillips S. Soil biota in an ungrazed invasion. Ecological Applications. 2001;**11**(5):1261-1275

[41] Shannon SM, Bauer JT, Anderson WE, Reynolds HL. Plant-soil feedbacks between invasive shrubs and native forest understory species lead to shifts in the abundance of mycorrhizal fungi. Plant and Soil. 2014;**382**:317-328

[42] Day NJ, Antunes PM, Dunfield KE. Changes in arbuscular mycorrhizal fungal communities during invasion by an exotic invasive plant. Acta Oecologica. 2015;**67**:66-74

[43] Jansa J, Smith FA, Smith SE. Are there benefits of simultaneous root colonization by different arbuscular mycorrhizal fungi? The New Phytologist. 2008;**177**:779-789

[44] Hausmann NT, Hawkes CV. Plant neighborhood control of arbuscular mycorrhizal community composition. The New Phytologist. 2009;**183**:1188-1200

[45] James TY, Kauff F, Schoch CL, Matheny PB, Hofstetter V, Cox CJ, et al. Reconstructing the early evolution of fungi using a six-gene phylogeny. Nature. 2006;**443**:818-822

[46] Read DJ, Perez-Moreno J. Mycorrhizas and nutrient cycling in ecosystems: A journey towards relevance? The New Phytologist. 2003;**157**:475-492

[47] Hobbie JE, Hobbie EA. 15N in symbiotic fungi and plants estimates nitrogen and carbon flux rates in arctic tundra. Ecology. 2006;**87**(4):816-822

[48] Wolfe BE, Rodgers VL, Stinson KA, Pringle A. The invasive plant *Alliaria petiolata* (garlic mustard) inhibits ectomycorrhizal fungi in its introduced range. Journal of Ecology. 2008;**96**:777-783

[49] Castellano SM, Gorchov DL. Reduced ectomycorrhizae on oak near invasive garlic mustard. Northeastern Naturalist. 2012;**19**:1-24

[50] Grove S, Haubensak KA, Parker IM. Direct and indirect effects of allelopathy in the soil legacy of an exotic plant invasion. Plant Ecology. 2012;**213**:1869-1882

[51] Sieverding E, Oehl F. Are arbuscular mycorrhizal fungal species invasive-derived from our knowledge about their distribution in different ecosystems? In: BCPC Symposium Proceedings. Plant Prot. Plant Heal. Eur. Introd. Spread Invasive Species. ResearchGate; 2005;**81**:197-202

[52] Hayward J, Horton TR, Pauchard A, Nunez MA. A single ectomycorrhizal fungal species can enable a Pinus invasion. Ecology. 2015;**96**:1438-1444

[53] Dickie IA, Nunez MA, Pringle A, Lebel T, Tourtellot SG, Johnston PR. Towards management of invasive

ectomycorrhizal fungi. Biological Invasions. 2016;**18**:3383-3395

[54] Van Der Putten WH. Plant defense belowground and spatiotemporal processes in natural vegetation. Ecology. 2003;**84**(9):2269-2280

[55] Bever JD, Westover KM, Antonovics J, Westover M. Incorporating the soil community into plant population dynamics: The utility of the feedback approach. Journal of Ecology. 1997;**85**(5):561-573

[56] Brasier C. Sudden oak death: *Phytophthora ramorum* exhibits transatlantic differences. Mycological Research. 2003;**107**(3):257-259

[57] Schwartz MW, Hoeksema JD, Gehring CA, Johnson NC, Klironomos JN, Abbott LK, et al. The promise and the potential consequences of the global transport of mycorrhizal fungal inoculum. Ecology Letters. 2006;**9**:501-515

[58] Weste G. Changes in the vegetation of sclerophyll shrubby woodland associated with invasion by *Phytophthova cinnamomi*. Australian Journal of Botany. 1981;**29**:261-276

[59] Shearer BL, Crane CE, Fairman RG, Grant MJ. Susceptibility of plant species in coastal dune vegetation of south-western Australia to killing by *Armillaria luteobubalina*. Australian Journal of Botany. 1998;**46**:321-334

[60] Venette RC, Cohen SD. Potential climatic suitability for establishment of *Phytophthora ramorum* within the contiguous United States. Forest Ecology and Management. 2006;**231**:18-26

[61] Bramble WC. Reaction of Chestnut bark to invasion by Endothia parasitica. American Journal of Botany. 1936;**23**(2):89-94

[62] Lovett GM, Canham CD, Arthur MA, Weathers KC, Fitzhugh RD. Forest ecosystem responses to exotic pests and pathogens in eastern North America. BioScience. 2006;**56**(5):395-405

[63] Pimentel D. Biological Invasions: Economic and Environmental Costs of Alien Plant, Animal, and Microbe Species. Ithaca: CRC Press; 2002

[64] Ellison AM, M.S. Bank, Clinton BD, Colburn EA, Elliott K, Ford CR, et al. Loss of foundation species: Consequences for the structure and dynamics of forested ecosystems. Frontiers in Ecology and the Environment. 2005;**3**(9):479-486

[65] Loo JA. Ecological impacts of non-indigenous invasive fungi as forest pathogens. Biological Invasions. 2009;**11**:81-96

[66] Liebhold AM, MacDonald WL, Bergdahl D, Mastro VC. Invasion by exotic forest pests: A threat to forest ecosystems. Forest Science. 1995;**30**:1-49

[67] Nijjer S, Rogers WE, Siemann E. Negative plant–soil feedbacks may limit persistence of an invasive tree due to rapid accumulation of soil pathogens. Proceedings of the Royal Society B: Biological Sciences. 2007;**274**:2621-2627

[68] Hol WHG, Van Veen JA. Pyrrolizidine alkaloids from Senecio jacobaea affect fungal growth. Journal of Chemical Ecology. 2002;**28**(9):1763-1772

[69] Van De Voorde TFJ, Van Der Putten WH, Bezemer TM. Intra- and interspecific plant–soil interactions, soil legacies and priority effects during old-field succession. Journal of Ecology. 2011:**99**;945-953

[70] Stinson KA, Campbell SA, Powell JR, Wolfe BE, Callaway RM, Thelen GC,

et al. Invasive plant suppresses the growth of native tree seedlings by disrupting belowground mutualisms. PLoS Biology. 2006;**4**(5):e140

[71] Eppinga MB, Rietkerk M, Dekker SC, De Ruiter PC. Accumulation of local pathogens: A new hypothesis to explain exotic plant invasions. Oikos. 2006;**114**(1):168-176

[72] Day NJ, Dunfield KE, Antunes PM. Fungi from a non-native invasive plant increase its growth but have different growth effects on native plants. Biological Invasions. 2016;**18**:231-243

[73] Rowe HI, Brown CS, Paschke MW. The influence of soil inoculum and nitrogen availability on restoration of high-elevation steppe communities invaded by *Bromus tectorum*. Restoration Ecology. 2009;**17**(5):686-694

[74] Requena N, Jimenez I, Toro M, Barea J. Interactions between plant-growth-promoting rhizobacteria (PGPR), arbuscular mycorrhizal fungi and rhizobium spp. in the rhizosphere of Anthyllis cytisoides, a model legume for revegetation in Mediterranean semi-arid ecosystems. The New Phytologist. 1997;**136**(4):667-677

[75] Mishra S, Chauhan PS, Goel AK, Upadhyay RS, Nautiyal CS. *Pseudomonas putida* NBRIC19 provides protection to neighboring plant diversity from invasive weed *Parthenium hysterophorus L*. by altering soil microbial community. Acta Physiologiae Plantarum. 2012;**34**(6):2187-2195

[76] Szili-Kovacs T, Torok K, Tilston EL, Hopkins DW. Promoting microbial immobilization of soil nitrogen during restoration of abandoned agricultural fields by organic additions. Biology and Fertility of Soils. 2007;**43**:823-828

[77] Lankau R. Soil microbial communities alter allelopathic competition between Alliaria petiolata and a native species. Biological Invasions. 2010;**12**(7):2059-2068

[78] Reinhart KO, Callaway RM. Soil biota and invasive plants. The New Phytologist. 2006;**170**(3):445-457

[79] Koziol L, Bever JD. The missing link in grassland restoration: Arbuscular mycorrhizal fungi inoculation increases plant diversity and accelerates succession. Journal of Applied Ecology. 2016;**54**:1301-1309

[80] Aprahamian AM, Lulow ME, Major MR, Balazs KR, Treseder KK, Maltz MR. Arbuscular mycorrhizal inoculation in coastal sage scrub. Botany. 2016;**499**:493-499

[81] Egli M, Mirabella a, Kägi B, Tomasone R, Colorio G. Influence of steam sterilisation on soil chemical characteristics, trace metals and clay mineralogy. Geoderma. 2006;**131** (1-2):123-142

[82] Jacobs DF. Toward development of silvical strategies for forest restoration of American chestnut (*Castanea dentata*) using blight-resistant hybrids. Biological Conservation. 2007;**137**:497-506

[83] Milgroom MG, Cortesi P. Biological control of Chestnut blight and hypovirulence: A critical analysis. Annual Review of Phytopathology. 2004;**42**:311-338

[84] Pliura A, Lygis V, Suchockas V, Bartkevicius E. Performance of twenty four European Fraxinus excelsior populations in three Lithuanian progeny trials with a special emphasis on resistance to Chalara fraxinea. Baltic Forestry. 2011;**17**(1):17-34

[85] Waring KM, Hara KLO. Silvicultural strategies in forest ecosystems affected by introduced pests. Forest Ecology and Management. 2005;**209**:27-41

[86] Dulmer KM, Leduc SD, Horton TR. Ectomycorrhizal inoculum potential of northeastern US forest soils for American chestnut restoration: Results from field and laboratory bioassays. Mycorrhiza. 2014;**24**:65-74

Chapter 4

Relationship between Exotic Plant Species Richness, Native Vegetation and Climate in Forest Ecosystems of Chile

Pablo I. Becerra and Ramiro O. Bustamante

Abstract

In this chapter, we evaluated how native species cover, native species richness, and foliage periodicity of forests are related to exotic species richness in different forest ecosystems distributed through the Mediterranean-type and temperate climate regions of Chile. We compiled published and unpublished data on Chilean vegetation (mainly phytosociological studies) located between 30° and 41° S in which plant species composition and abundance (percentage of cover) per plot was evaluated and published. When all forests were analysed together, we found that exotic species richness was not significantly related to native species richness; it was negatively related to native species cover in both Mediterranean and temperate climatic regions and was greater in deciduous than evergreen forests, although only in the temperate region. In an analysis considering separately every forest type, we observed that only in 50% of these communities, exotic species richness was negatively related to native species cover. Furthermore, in 25% of forest types, exotic species richness was negatively related to native species richness, and in only one forest type, this relationship was positive. Our results suggest that native species produce some resistance against invasion of exotic plants, but this effect may depend on climate and forest type.

Keywords: alien species, biotic resistance, species diversity, forest communities, plant invasion

1. Introduction

Given the impact of invasive species on native biodiversity, it is important to increase the knowledge about factors influencing invasion processes to establish conservation and restoration strategies [1–3]. Invasion by exotic plant species has been related to biotic (native vegetation and herbivory) as well as abiotic (e.g. climate and soil) factors [3, 4–7]. Among biotic factors, native vegetation has received more attention. Native vegetation may negatively affect invasion of exotic plants through competitive interactions as native species use resources that exotic species also require (space, light, nutrients, etc.) [2, 7, 8]. However, positive effects and facilitation of native species on exotic plants have also been observed [9]. Amelioration of abiotic stress by native vegetation may facilitate recruitment and/or establishment of exotic species, especially under more stressful conditions [10–12]. As a consequence of these negative or positive interactions, different relationships between native vegetation variables, such as species richness and cover, and invasibility may be expected. For instance, richer communities should reduce niche opportunities for establishing exotic species and hence resist invasion better than poorer communities, and in this case, a negative relationship between native and exotic richness may be expected [2, 7, 13, 14]. This pattern should occur in the absence of covarying extrinsic biotic or abiotic factors [15], which would mainly occur at small spatial scales or within community types [7]. Instead, positive relationships between exotic and native species richness should occur when competition is less important than facilitation [16, 17] or when both, native and exotic species, respond similarly to external environmental factors, which would mainly occur in large spatial scales often including different communities and climates [7]. Empirical evidence shows different results: negative and positive as well as absence of significant relationships have been documented [18–26] although negative relationships have mainly been observed in local scales while positive relationships at larger spatial scales.

Similarly, reductions in native species cover (e.g. by disturbance) would increase plant invasion through depletion of competitive interactions [13, 27]. Many studies have documented that more disturbed habitats, where cover of native plants has been eliminated or strongly reduced, present higher plant invasion [4, 5, 18–20, 25, 28–31]. Nonetheless, it is less clear if within forest habitats, sites with lower cover of native species (e.g. triggered by some natural or anthropogenic disturbance) but without complete elimination of the forest canopy are related to an increase in plant invasion. Alternatively, when facilitative interactions are stronger than competition, positive associations between native and exotic species are more frequent [11], and positive relationships between native species cover and plant invasion may be expected. Moreover, some studies have recently documented invasion in closed-canopy forests by some shade-tolerant exotic plant species [32], and hence for some of these exotic species, greater cover of native species may even be necessary.

On the other hand, the effect of native vegetation on plant invasion may be related to other variables. An attribute of native vegetation, which may affect plant invasion, is the foliage periodicity of dominant tree species (e.g. evergreen and deciduous) [33]. Forest communities dominated by deciduous species receive more light than evergreen forests, at least during a period of time within a year, and thus, evergreen forests may produce greater competitive resistance against plant invasion. However, the role of this factor in plant invasion has been very few times evaluated. Only Ibáñez et al. [33] addressed this issue and documented greater exotic species richness in deciduous forests than evergreen forests in a region of eastern North America.

Plant-plant interactions may strongly be modulated by abiotic conditions [34], and hence, the effect of native vegetation on plant invasion may be influenced by climate or habitat conditions. Specifically, positive interactions should be more frequent under more stressing conditions, while competition under more productive or favourable conditions for plants, which is known as the stress-gradient hypothesis [34]. Although some evidence for this hypothesis has been documented through experimental approaches in the context of plant invasions [10–12, 17], these studies have only been performed at local scales and for particular exotic species. Thus, it is less known that the pattern of relationship between native vegetation variables, such as cover and species richness, and exotic invasion varies between different climatic or habitat conditions according to the stress-gradient hypothesis (but see [25]). Based on the stress-gradient hypothesis, plant invasion should be positively related to cover and richness of native species under more stressing conditions (e.g. arid and semiarid climates), while negatively in more favourable conditions (e.g. temperate climates). Likewise, plant invasion should be greater in deciduous forests than evergreen forests under more favourable conditions.

In this study, we compiled data on exotic species richness at a plot scale in different forest types of Chile and assessed how exotic species richness is correlated to cover and richness of native species, as well as the type of periodicity of foliage of dominant trees in forests distributed through the Mediterranean-type and temperate regions of the country.

2. Methods

2.1 Data source

We compiled several published and unpublished studies on Chilean vegetation, from which we obtained data on exotic and native species composition and abundance at a plot scale from different types of forest communities (**Table 1**). Each one of these communities corresponded to a particular phytosociological unit, most of them at the association level, with homogeneous climate and soil conditions. However, due to differences in geographical distribution among forest types, these had different climatic regimes (Mediterranean-type and temperate climates) and are dominated by species with different periodicity of foliage (deciduous or evergreen). We considered only studies in which species composition and abundance (percentage of cover) per plot were published (e.g. phytosociological tables), from which we obtained values of exotic and native species richness and cover. Within every forest type, selected plots for the analyses had the same size, although the size of plots varied between some forest types. In addition, the distance between plots varied between forest types, from 300 m in some forest types to some kilometres in others. Also, forest types included in the analyses did not show significant evidence (indicated in the publication) of strong anthropogenic disturbances (massive logging of trees or recent fires). This entailed that all plots within each forest types were at least partially covered by a tree canopy. However, forest types could present slight perturbations such as herbivory by exotic livestock and rabbits, tourism impacts (e.g. tracks) and/or reduced past forest logging. This produced a gradient of native species cover and richness within all forests. According to these criteria, we found eight forest types including Mediterranean as well as temperate forests of Chile (**Table 1**).

2.2 Forest types selected for the study

Forest types selected for this study were distributed between 33 and 41°S of Chile, covering both Mediterranean and temperate climates (**Table 1** and **Figure 1**). The most septentrional forest used in this study is a sclerophyllous forest (Becerra, unpublished data), which is an evergreen forest distributed in the coastal range of the semi-arid Mediterranean zone, dominated by the species *Cryptocarya alba* and *Schinus velutina*. The Mediterranean montane deciduous forest [35] is dominated by the deciduous species *Nothofagus glauca* and is distributed in coastal as well as pre-Andean foothills of the Mediterranean region. The Mediterranean Subantarctic Andean forest [36] is dominated by the deciduous species *Nothofagus pumilio* and is distributed along the Andean timberline within the humid Mediterranean region of Chile. The temperate montane deciduous forest [37] is dominated by the deciduous species *Nothofagus alpina* and *N. obliqua* and is distributed on the foothills of the Andean range. The swamp deciduous forest [38] is dominated by the deciduous species *Nothofagus antarctica* and is distributed on swamp soils in the central valley of the temperate region. The lowland evergreen forest [38] is dominated by the evergreen species *Eucryphia cordifolia* and *Nothofagus dombeyi* and is distributed on the foothills of the Andean range of

Forest type	Lat.	Long.	Plot (m^2)	N° plots	Region	Foliage	% herbs	% shrubs	% trees
Sclerophyllous forest (Sc)	32°17′	71°11′	100	18	M	E	100	0.0	0.0
Mediterranean montane deciduous forest (Mm)	35–36°	71–72°	100	21	M	D	28.6	42.9	28.6
Mediterranean Subantarctic Andean forest (Sa)	35°35′	71°02′	100	26	M	D	83.3	16.7	0.0
Temperate montane deciduous forest (Tm)	38°26′	71°31′	100	20	T	E	95.0	5.0	0.0
Lowland deciduous forest (Ld)	39–40°	72–73°	140	35	T	D	88.0	12.0	0.0
Swamp evergreen forest (Se)	39–41°	72–73°	100	27	T	E	85.7	7.1	7.1
Swamp deciduous forest (Sd)	40°30′	72°30′	200	10	T	D	90.0	10.0	0.0
Lowland evergreen forest (Le)	41°00′	73°00′	200	10	T	E	90.5	9.5	0.0

Table 1.
Characteristics of the studied forest ecosystems: geographical location or range in which each forest was sampled (latitude and longitude), size of sampling plots, number of plots used, climatic region (M: Mediterranean; T: temperate), foliage periodicity of dominant species (E: evergreen; D: deciduous) and percentages of herb, shrub and tree species in the exotic flora in each forest.

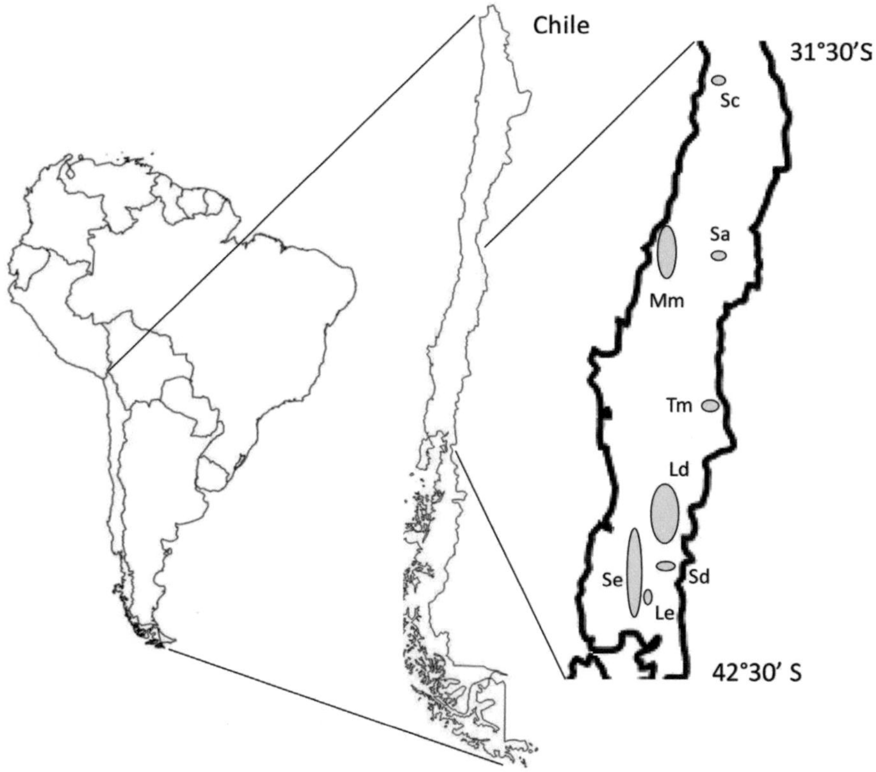

Figure 1.
Geographical distribution (only referential location) of forest types included in the study. Different studies covered different surfaces. Nomenclature of forest types is indicated in **Table 1**.

the temperate region. The lowland deciduous forest [39] is dominated by the deciduous species *Nothofagus obliqua* and is distributed in the central valley of the temperate region. Finally, the swamp evergreen forest [40] is dominated by the evergreen species *Myrceugenia exsucca* and *Blepharocalyx cruckshanksii* and is distributed on the lowlands of the temperate region of Chile occupying mainly riverine habitats.

Climate of forests included in this study varied from approximately 350 mm of annual precipitation and an annual mean temperature of 14°C in the sclerophyllous forest, to near 2500 mm of annual precipitation and an annual mean temperature of 10.5°C in the lowland evergreen forest [41].

2.3 Data analysis

We performed two types of analyses: first, we analysed together all data from all forest types, in order to examine the contribution of every variable to the variation in exotic species richness. Thus, in the same model, we evaluated the independent effects of the climatic region (Mediterranean vs. temperate), foliage periodicity of forests (deciduous vs. evergreen), native species richness and native species cover, as well as the statistical interactions between the climatic region and each native vegetation variable. Although the areas of plots were in general quite similar between forests (**Table 1**), the size of plots differed between some of them (**Table 1**), and preliminary $\log_{(10)}$ area-log (N° species) correlations were significant for all forest types. Hence, to compare exotic and native species richness at a plot scale among all forest types, we controlled the size of plots by dividing values of exotic species richness as well as native species richness by the $\log_{(10)}$ of the area of each plot. On the other hand, cover values of native species per plot were quantified by the sum of cover among all native species per plot. However, due to larger plots that may have more species and therefore more components for this sum, to use native species cover in the analyses, we controlled the differences in the species number by dividing the sum of cover by the number of native species per plot. Thus, we obtained a variable representing a mean cover of native species per plot.

Finally, we assessed the relationships between exotic species richness and native species richness and cover separately for each forest type with the aim to evaluate if these intra-forest relationships are generalised among different forest types in central-south Chile. In these cases, because within each forest type the size of plots was equal among plots, we did not control the area of them and used the absolute number of exotic and native species as well as the sum of native species cover directly.

All statistical analyses were carried out using SPSS 15.0 by generalised linear models (GLMs).

3. Results

3.1 Exotic flora

Among the eight forest types, we recorded 56 exotic species corresponding to three trees (5.4%), four shrubs (7.1%) and 49 herbs (87.5%). The most common species were *Rumex acetosella* and *Rosa rubiginosa* present in six forest types; *Hypochaeris radicata*, *Prunella vulgaris* and *Veronica serpyllifolia* present in five forest types; and *Anthoxanthum odoratum*, *Holcus lanatus*, *Leontodon taraxacoides*, *Lotus uliginosus*, *Plantago lanceolata*, *Trifolium repens* and *Rubus constrictus* present in four forest types. In general, exotic species were mostly herbs in all forests with percentages greater than 80%, except in the Mediterranean deciduous forest (**Table 1**). Species composition per forest-type is shown in **Table 2**.

3.2 Relationship between native vegetation and exotic species richness

Regarding all forests, exotic species richness varied between 0 and 20 species per plot. After controlling the area of plots, exotic species richness was significantly greater in temperate forests than in Mediterranean forests along all gradients of cover and richness of native species and for each type of forest canopy (deciduous or evergreen) (**Table 3** and **Figures 2–4**). Exotic species richness was negatively and significantly related to the native species cover (**Table 3** and **Figure 2**). This pattern seems to occur in both Mediterranean and temperate forests as we found no significant interaction between the climatic region and native species cover (**Table 3**). However, the slope of this relationship in plots from the temperate region was greater than in plots from the Mediterranean-type climate region (**Figure 2**). In contrast, exotic species richness was not significantly related to the native species richness when all plots were analysed together or in each climatic region separately (**Table 3** and **Figure 3**). On the other hand, exotic species richness was significantly greater in deciduous forests than evergreen forests (**Table 3** and **Figure 4**). Yet, the interaction between region and foliage periodicity was significant (**Table 3**), indicating that higher exotic species richness in deciduous forests than evergreen forests occurred only in temperate forests (**Figure 4**).

When analysing data separately for each forest type, they showed different relationships between native vegetation variables and exotic species richness. A significant negative relationship between native species cover and exotic species

Exotic species	Sc	Mm	Sa	Tm	Sd	Le	Ld	Se
Achillea millefolium				1				
Agrostis capillaris		1		1			1	
Agrostis castellana					1	1		
Agrostis tenuis								1
Anagallis arvensis					1			
Anthoxanthum odoratum				1	1	1	1	
Aster vahlii					1			
Bellardia trixago					1	1		
Bromus hordeaceus			1					
Capsella bursa-pastoris	1							
Cirsium vulgare							1	1
Crataegus monogyna		1						
Crepis capillaris				1				
Cynosurus echinatus				1			1	
Cytisus striatus		1						
Chrysanthemum sp.			1					
Dactylis glomerata				1		1	1	
Digitalis purpurea						1	1	1
Erodium cicutarium	1							
Fumaria officinalis	1							
Galium aparine	1							
Gastridium ventricosum			1					
Holcus lanatus					1	1	1	1
Hypochoeris radicata				1	1	1	1	1
Lapsana communis				1				
Leontodon taraxacoides				1	1	1	1	
Leucanthemum vulgare						1	1	
Lolium multiflorum							1	
Lolium perenne					1		1	
Lotus uliginosus					1	1	1	1
Medicago polymorpha								1
Mentha pulegium								1
Panicum capillare							1	
Phleum pratense				1				
Pinus radiata		1						
Plantago lanceolata				1	1	1	1	
Poa pratensis				1				
Poa trivialis				1				
Prunella vulgaris				1	1	1	1	1
Ranunculus repens					1	1		
Rosa rubiginosa		1	1	1	1	1	1	
Rubus constrictus					1	1	1	1
Rubus ulmifolius		1						
Rumex acetosella			1	1	1	1	1	1
Salix viminalis								1
Sonchus asper							1	
Stellaria media	1					1		
Taraxacum officinale				1		1	1	

Exotic species	Sc	Mm	Sa	Tm	Sd	Le	Ld	Se
Teline monspessulana		1					1	
Trifolium dubium					1		1	
Trifolium pratense	1					1	1	
Trifolium repens				1	1		1	1
Urtica dioica						1		
Verbascum thapsus				1				
Veronica scutellata					1			
Veronica serpyllifolia			1	1	1	1		1

Table 2.
Exotic species recorded in each forest type. Nomenclature of forests is in **Table 1.**

Source of variation	Chi^2	P
Climatic region	54.999	<0.001
Foliage periodicity	8.377	0.004
Native cover	5.201	0.023
Native richness	2.241	0.134
Climatic region × foliage	17.278	<0.001
Climatic region × native cover	0.029	0.864
Climatic region × native richness	0.066	0.798

Table 3.
Statistical results (generalised lineal models) for the effect of climatic region (Mediterranean vs. Temperate), foliage periodicity (evergreen vs. deciduous), native species cover and native species richness on exotic species richness (N = 167 plots).

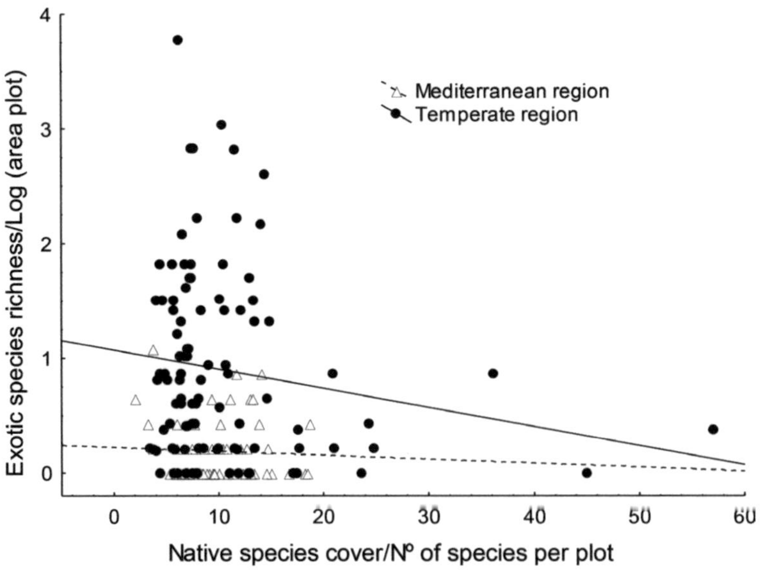

Figure 2.
Relationship between exotic species richness and native species cover per climatic region. Exotic species richness per plot was divided by the logarithm (10) of the area of plots to control the differences of area between plots. Native species cover per plot was divided by the number of native species in the plot to control the differences of richness between plots.

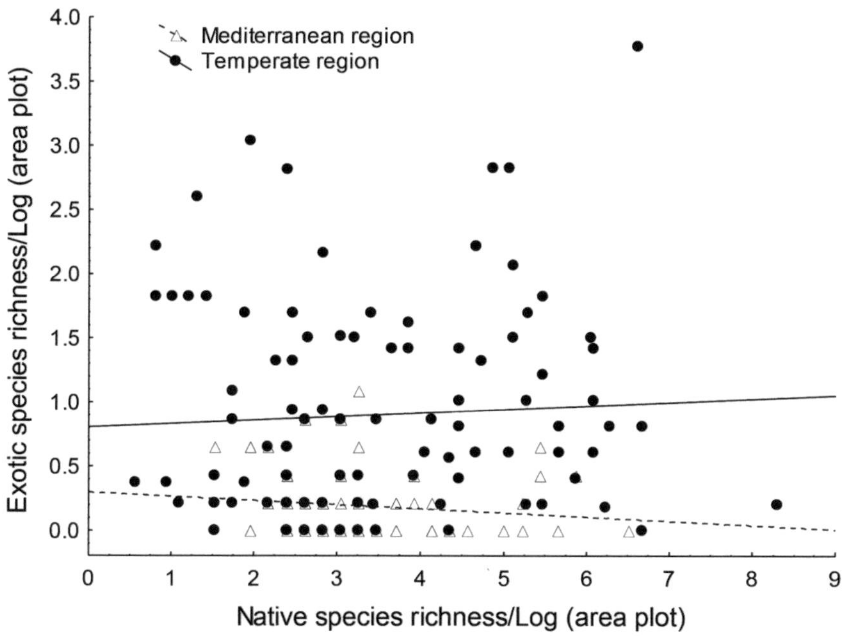

Figure 3.
Relationship between exotic species richness and native species richness per climatic region. Exotic species richness and native species richness per plot were divided by the logarithm (10) of the area of plots to control the differences of area between plots.

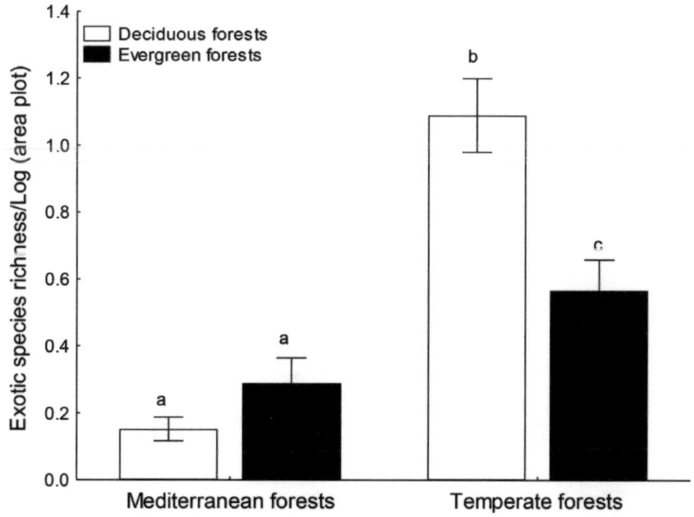

Figure 4.
Exotic species richness per climatic region and foliage periodicity type (mean ± 1 S.E.) (N = 167 plots). Different letters indicate significant statistical differences between each combination of region and foliage periodicity (P < 0.05).

richness was observed in four forest types, representing 50% of studied forest types, two forests from the Mediterranean-type climate region and two from the temperate region (**Table 4** and **Figure 5**). There was no positive relationship between native species cover and exotic species richness. In turn, a significant relationship between native species richness and exotic species richness was observed only in three forest types, representing 38% of forests included in this study, all of them corresponding

Forest type	Native species richness		Total native cover	
	Chi^2	P	Chi^2	P
Sclerophyllous forest (Sc)	1.901	0.168	0.626	0.429
Mediterranean Subantarctic Andean forest (Sa)	1.217	0.270	7.337	**0.007**
Mediterranean montane deciduous forest (Mm)	1.043	0.307	8.109	**0.004**
Lowland deciduous forest (Ld)	11.351	**0.001**	0.012	0.913
Swamp deciduous forest (Sd)	7.781	**0.005**	0.550	0.458
Lowland evergreen forest (Le)	0.074	0.786	0.058	0.810
Temperate montane deciduous forest (Tm)	4.526	**0.033**	11.450	**0.001**
Swamp evergreen forest (Se)	0.004	0.951	6.160	**0.013**

Table 4.
Statistical results (generalised linear models, ordinal multinomial distribution of data and logit function link) of analyses per forest type for the effect of native species richness and native species cover on exotic species richness. Significant relationships in bold.

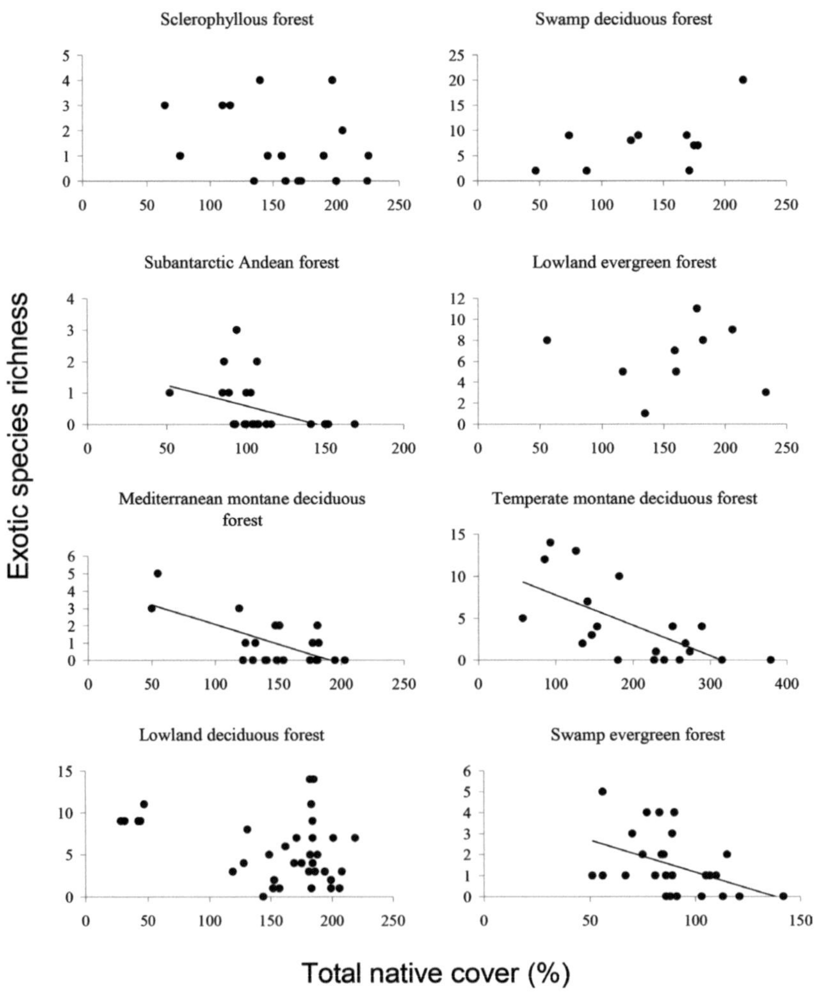

Figure 5.
Relationships between exotic species richness and native species cover in every forest type. Curves are shown only for significant relationships.

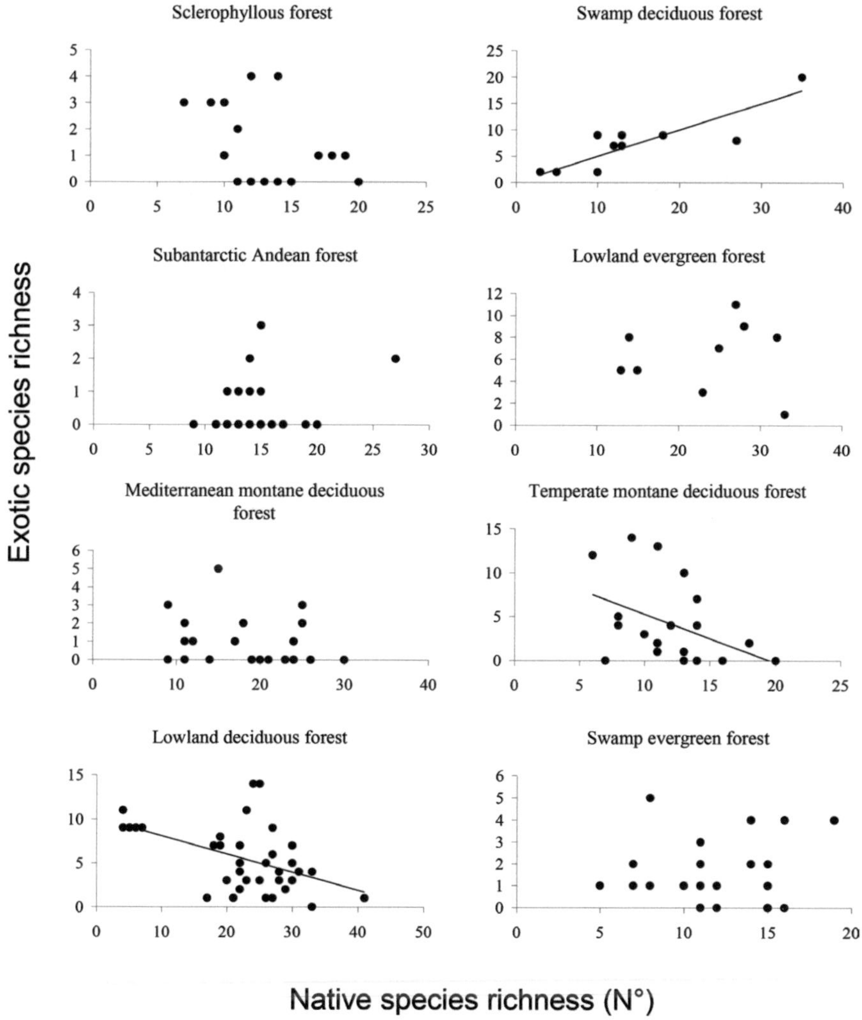

Figure 6.
Relationships between exotic species richness and native species richness in every forest type. Curves are shown only for significant relationships.

to temperate forests (**Table 4** and **Figure 6**). However, in this case, in two forests the relationship was negative while in one forest type (Swamp deciduous forest) the relationship was positive (**Figure 6**).

4. Discussion

In this study, we document that exotic species richness is related to variation in native species cover, foliage periodicity and, at a less extent, native species richness. Additionally, these relationships depend on climate and/or forest type in forest communities of Chile.

The significant negative relationships between exotic species richness and native species cover pooling all forests as well as within some forest types suggest competitive effects of native vegetation on invasion of exotic plants [27]. Variability of native cover within forests may be produced by natural causes (including natural disturbances) as well as anthropogenic disturbances. Regardless of the cause determining this variability, lower native cover entails more resources for invasive species [13, 14, 27]. Therefore, our results suggest a high importance of competition in invasion processes of exotic species in these forest communities. Similarly, other observational studies performed within forest ecosystems [19, 28, 31, 42] as well as an increasing number of experimental studies have demonstrated the importance of resource availability and competition liberation for plant invasion [8, 12, 21, 23, 43]. Therefore, although many exotic plant species may also invade closed-canopy forests, as documented by Martin et al. [32], our results suggest that more covered sites within or between native forests may better resist plant invasion.

Globally within the study area, we observed no significant relationship between exotic and native species richness. Instead, when every forest was separately analysed, two of them showed a significant negative relationship, which is consistent with the idea that negative relationships between exotic and native species richness would only occur when other factors are controlled (i.e. within each forest type) [3, 7, 15, 21]. This result suggests that, at least in these two forest types, exotic and native species may be competing by resources. However, our results did not agree to Shea and Chesson [7] and some empirical studies [6, 19, 22, 25, 44, 45], which proposed that in geographical comparisons (in our case in the analysis pooling data from all forests), positive relationships between exotic and native species richness should emerge. In particular, the absence of a positive correlation between native and exotic species richness when all forests were analysed together contrasts to Fuentes et al. [26], who found similar geographical tendencies between native and exotic species richness along Chile, although in this case, using much larger scales to measure species richness. Davies et al. [46] proposed that positive relationships between exotic and native species richness would mainly occur when richness is quantified at large spatial scales [26]. This would occur because greater environmental heterogeneity within large quadrants would favour both exotic and native species. Instead, this would not occur at small spatial scales such as a plot-scale (this study), even though plots are compared at a geographical scale [46]. However, Souza et al. [47] found that native and exotic species richness may be positively correlated both at local and landscape scales. Consistent to Souza et al. [47], we observed a positive relationship between exotic and native species richness occurring within a forest type (at a local scale). This positive correlation could be produced by a similar response of native and exotic species to environmental factors [7, 47], or because native species are facilitating exotic species [16, 47]. In the forest type in which this positive relationship was observed (swamp deciduous forest), soil conditions are extreme, with soils permanently saturated with water and low soil oxygen [38]. Therefore, it is probable that within this community, only in microsites (at a scale of 100 m^2 or less) where soil conditions are a little more favourable, more species, exotics as well as natives, can coexist. Nevertheless, it is not possible to rule out facilitative effects of native on exotic species in this case either.

We observed that in the temperate region, deciduous forests presented greater exotic species richness than evergreen forests, which suggests that the seasonal increase in light conditions in deciduous forests may be a factor contributing to an increase of invasion probability in temperate forests. Higher light requirement of exotic species has been proposed as an important life history attribute favouring the invasion in low-cover sites [14, 43], for instance, ruderal habitats. In fact, most of the exotic species of central-south of Chile have been documented as shade-intolerant species [48]. Although the relationship between foliage periodicity and plant invasion has scarcely been evaluated, our results agree to Ibáñez et al. [33] who documented greater exotic species richness in deciduous forests than evergreen forests in eastern North America. Thus, this factor may be an important driver of plant invasion in forest ecosystems although more studies are needed to assess the generality of this relationship.

On the other hand, our results show that under the same conditions of cover, richness and foliage periodicity of native species, forests from the temperate climate region were richer in exotic species than forests from the Mediterranean-type climate region. This suggests that the Mediterranean-type semiarid region of Chile represents a more stressful condition than the temperate region for exotic species invading forest ecosystems, which agrees to several studies documenting that greater exotic species richness seems to establish mainly in more productive climates [6, 20, 26, 33]. For instance, Lonsdale [6] found a lower number of exotic species in deserts and savannas than in forest habitats around the world, and Stohlgren et al. [20] found a positive relationship between productivity and exotic species richness within North America. Ibáñez et al. [33] documented higher exotic species richness in areas with warmer temperatures and higher summer precipitation in a region of eastern North America. Finally, in the same country, Chile, Fuentes et al. [26] observed greater exotic species richness at a regional scale (10 × 10 km) within Mediterranean and temperate regions than in deserts or colder areas. Nevertheless, in contrast to our results, Fuentes et al. [26] observed higher exotic species richness in the Mediterranean region than the temperate region although, in this case, using larger scales to measure exotic species richness.

If more xeric climates such as in the Mediterranean region of Chile entail more stressful conditions for exotic species (as suggested by the greater exotic species richness in the temperate climate), based on the stress-gradient hypothesis [34], in the Mediterranean region, facilitative interactions and positive relationships may be expected between native vegetation and exotic species richness, while competitive and negative relationships in temperate forests. Thus, the fact that the relationship between native cover and plant invasion was observed in both climatic regions only partially supports the stress-gradient hypothesis [34]. However, the slope of this relationship was steeper in the temperate region than in the Mediterranean-type climate region, suggesting that competitive effects of native vegetation on exotic species were stronger in the less stressful climatic region, which agrees to the stress-gradient hypothesis [11, 34]. Additionally, we did not find a significant negative relationship between native cover and exotic species richness in the most xeric forest included in this study (sclerophyllous forest), which again suggests that under more stressing conditions in terms of water availability, competition would be weaker, or that facilitative effects of native vegetation on exotic species counteracted any competitive interaction (e.g. [10, 12, 17]). Furthermore, two among five temperate forests presented negative relationships between native cover and exotic richness, and in the other temperate forest (lowland deciduous forest), a negative relationship between native and exotic species richness was observed. These results suggest that in temperate forests negative relationships between native vegetation and exotic species are more frequent than in Mediterranean forests, which agree with the stress-gradient hypothesis. On the other hand, the fact that deciduous forests presented greater exotic richness than evergreen forests only in the temperate region suggests that in more humid regions the light conditions may be a more important limiting factor for exotic species than in Mediterranean-type climates. This result may also be consistent with the stress-gradient hypothesis since competition by light would be stronger under less stressful abiotic conditions (temperate region). In consequence, our results suggest that the stress-gradient hypothesis may be useful to predict patterns of relationship between exotic species richness and native vegetation when species richness is analysed at small spatial scales. Other studies [11] have also found support for this hypothesis in the context of interactions between native and exotic species.

Finally, our results suggest that maintaining or increasing native species cover may help to control or reduce plant invasion, at least in terms of exotic species richness. This may be a successful management strategy for the control of invasion mainly in temperate forests as well as in some Mediterranean-type climate forests. Instead, in more xeric Mediterranean forests (e.g. sclerophyllous forest), an increase in cover of native species does not seem to be enough to reduce exotic species richness, and other actions are needed to control plant invasion.

Acknowledgements

We thank the Chilean phytosociologists who have greatly contributed to knowledge of Chilean plant communities and have published original tables of species which allowed performing this study. This work was supported by a doctoral fellowship from CONICYT to PIB and by ICM-P05-002. PIB thanks FB 0002-2014.

Author details

Pablo I. Becerra[1,2]* and Ramiro O. Bustamante[3,4]

1 Departamento de Ecosistemas y Medio Ambiente, Facultad de Agronomía e Ingeniería Forestal Pontificia Universidad Católica de Chile, Santiago, Chile

2 Center of Applied Ecology and Sustainability, CAPES, Santiago, Chile

3 Instituto de Ecología y Biodiversidad, Santiago, Chile

4 Departamento de Ciencias Ecológicas, Facultad de Ciencias, Universidad de Chile, Santiago, Chile

*Address all correspondence to: pablobecerra@uc.cl

References

[1] Groves R, Burdon J. Ecology of Biological Invasions. Cambridge, UK: Cambridge University Press; 1986. 180 p

[2] Drake J, Mooney H, di Castri F, Groves R, Kruger F, Rejmanek M, et al. Biological Invasions: A Global Perspective. Chichester, UK: Wiley; 1989. 550 p

[3] Huston MA. Management strategies for plant invasions: Manipulating productivity, disturbance, and competition. Diversity and Distributions. 2004;**10**:167-178

[4] Jauni M, Gripenberg S, Ramula S. Non-native plant species benefit from disturbance: A meta-analysis. Oikos. 2015;**124**:122-129. DOI: 10.1111/oik.01416

[5] Catford J, Daehler C, Murphy H, Sheppard A, Hardesty B, Westcott D, et al. The intermediate disturbance hypothesis and plant invasions: Implications for species richness and management. Perspectives in Plant Ecology, Evolution and Systematics. 2012;**14**:231-241. DOI: 10.1016/j.ppees.2011.12.002

[6] Lonsdale W. Global patterns of plant invasions and the concept of invasibility. Ecology. 1999;**80**:1522-1536

[7] Shea K, Chesson P. Community ecology theory as a framework for biological invasions. Trends in Ecology & Evolution. 2002;**17**:170-176

[8] Davis MA, Pelsor M. Experimental support for a resource-based mechanistic model of invasibility. Ecology Letters. 2001;**4**:421-428

[9] Bruno JF, Fridley JD, Bromberg KD, Bertness MD. Insights into biotic interactions from studies of species invasions. In: Sax DF, Stachowicz JJ, Gaines SD, editors. Species Invasions: Insights into Ecology, Evolution and Biogeography. USA: Sinauer Associates; 2005. pp. 13-40

[10] Lenz TI, Facelli JM. Shade facilitates an invasive stem succulent in a chenopod shrubland in South Australia. Austral Ecology. 2003;**28**:480-490

[11] Von Holle B. Environmental stress alters native-nonnative relationships at the community scale. Biological Invasions. 2013;**15**:417-427. DOI: 10.1007/s10530-012-0297-7

[12] Becerra P, Bustamante R. Effect of a native tree on seedling establishment of two exotic species in a semiarid ecosystem. Biological Invasions. 2011;**13**:2763-2773

[13] Hobbs R, Huenneke L. Disturbance, diversity, and invasions: Implications for conservation. Conservation Biology. 1992;**6**:324-337

[14] Bartomeus I, Sol D, Pino J, Vicente P, Font X. Deconstructing the native-exotic richness relationship in plants. Global Ecology and Biogeography. 2012;**21**:524-533. DOI: 10.1111/j.1466-8238.2011.00708.x

[15] Naeem S, Knops J, Tilman D, Howe K, Kennedy T, Gale S. Plant diversity increases resistance to invasion in the absence of covarying extrinsic factors. Oikos. 2000;**91**:97-108

[16] Bruno JF, Stachowicz JJ, Bertness MD. Inclusion of facilitation into ecological theory. Trends in Ecology & Evolution. 2003;**18**:119-125

[17] Von Holle B. Biotic resistance to invader establishment of a southern Appalachian plant community is determined by environmental conditions. Journal of Ecology. 2005;**93**:16-26

[18] Planty-Tabacchi A, Tabacchi E, Naiman R, DeFerrari C, DéCamps H. Invasibility of species-rich communities in riparian zones. Conservation Biology. 1996;**10**:598-607

[19] Stohlgren TJ, Binkley D, Chong G, Kalkhan M, Schell L, Bull K, et al. Exotic plant species invade hot spots of native plant diversity. Ecological Monographs. 1999;**69**:25-46

[20] Stohlgren TJ, Barnett D, Flather C, Kartesz J, Peterjohn B. Plant species invasions along the latitudinal gradient in the United States. Ecology. 2005;**86**:2298-2309

[21] Levine J. Local interactions, dispersal, and native and exotic plant diversity along a California stream. Oikos. 2001;**95**:397-408

[22] Espinosa-García FJ, Villaseñor JL, Vibrans H. The rich generally get richer, but there are exceptions: Correlations between species richness of native plant species and exotic weeds in Mexico. Diversity and Distributions. 2004;**10**:399-407

[23] Bruno JF, Kennedy CW, Rand TA, Grant M. Landscape-scale patterns of biological invasions in shoreline plant communities. Oikos. 2004;**107**:531-540

[24] Howard TG, Gurevitch J, Hyatt L, Carreiro M, Lerdau M. Forest invasibility in communities in southeastern New York. Biological Invasions. 2004;**6**:393-410

[25] Perelman SB, Chaneton EJ, Batista WB, Burkart SE, León JC. Habitat stress, species pool size and biotic resistance influence exotic plant richness in the flooding Pampa grasslands. Journal of Ecology. 2007;**95**:662-673

[26] Fuentes N, Pauchard A, Sánchez P, Esquivel J, Marticorena A. A new comprehensive database of alien plant species in Chile based on herbarium records. Biological Invasions. 2013;**15**:847-858. DOI: 10.1007/s10530-012-0334-6

[27] Davis MA, Grime JP, Thompson K. Fluctuating resources in plant communities: A general theory of invasibility. Journal of Ecology. 2000;**88**:528-534

[28] Halpern C, Spies T. Plant species diversity in natural and managed forests of the pacific northwest. Ecological Applications. 1995;**5**:913-934

[29] Wiser S, Allen R, Clinton P, Platt K. Community structure and forest invasion by an exotic herb over 23 years. Ecology. 1998;**79**:2071-2081

[30] Simberloff D, Relva MA, Nuñez M. Gringos en el bosque: Introduced tree invasion in a native *Nothofagus/Austrocedrus* forest. Biological Invasions. 2002;**4**:35-53

[31] Aragón R, Morales JM. Species composition and invasion in NW Argentinian secondary forests: Effects of land use history, environment and landscape. Journal of Vegetation Science. 2003;**14**:195-204

[32] Martin PH, Canham CD, Marks PL. Why forests appear resistant to exotic plant invasions: Intentional introductions, stand dynamics, and the role of shade tolerance. Frontiers in Ecology and the Environment. 2009;**7**:142-149

[33] Ibáñez I, Silander JA, Allen JM, Treanor SA, Wilson A. Identifying hotspots for plant invasions and forescating focal points of further spread. Journal of Applied Ecology. 2009;**46**:1219-1228

[34] Maestre FT, Callaway RM, Valladares F, Lortie CJ. Refining the stress-gradient hypothesis for competition and facilitation plant communities. Journal of Ecology. 2009;**97**:199-205

[35] Amigo J, San Martin J, García L. Estudio fitosociológico de los bosques de *Nothofagus glauca* (Phil.) Krasser del centro-sur de Chile. Phytocoenologia. 2000;**30**:193-221

[36] San Martín J, Troncoso A, Mesa A, Bravo T, Ramírez C. Estudio fitosociológico del bosque caducifolio magallánico en el límite norte de su área de distribución. Bosque. 1991;**12**:29-41

[37] Becerra P, Cruz G. Diversidad vegetacional de la Reserva Nacional Malalcahuello, IX región de Chile. Bosque. 2000;**21**:47-68

[38] Ferrada V. Estudio fitosociológico del Ñadi de Frutillar (Osorno, Chile) [thesis]. Facultad de Ciencias Forestales: Universidad Austral de Chile; 1987

[39] San Martín C, Ramírez C, Figueroa H, Ojeda N. Estudio sinecológico del bosque de roble-laurel-lingue del centro sur de Chile. Bosque. 1991;**12**:11-27

[40] Ramírez C, Ferriere F, Figueroa H. Estudio fitosociológico de los bosques pantanosos templados del sur de Chile. Revista Chilena de Historia Natural. 1983;**56**:11-26

[41] Luebert F, Pliscoff P. Sinopsis bioclimática y vegetacional de Chile. Santiago, Chile: Editorial Universitaria. 316 p

[42] Teo D, Tan H, Corlett R, Min Wong C, Lum S. Continental rain forest fragments in Singapore resist invasion by exotic plants. Journal of Biogeography. 2003;**30**:305-310

[43] Grotkopp EM, Rejmánek M, Rost TL. Toward a causal explanation of plant invasiveness: Seedling growth and life-history strategies of 29 pine (*Pinus*) species. The American Naturalist. 2002;**159**:396-419

[44] Chaneton E, Perelman S, Omacini M, León R. Grazing, environmental heterogeneity and exotic invasions in temperate Pampa grasslands. Biological Invasions. 2002;**4**:7-24

[45] Sax DF. Native and naturalized plant diversity are positively correlated in scrub communities in California and Chile. Diversity and Distributions. 2002;**8**:193-210

[46] Davies KF, Chesson P, Harrison S, Inouye BD, Melbourne BA, Rice KJ. Spatial heterogeneity explains the scale dependence of the native-exotic diversity relationship. Ecology. 2005;**86**:1602-1610

[47] Souza L, Bunn W, Simberloff D, Lawton R, Sanders N. Biotic and abiotic influences on native and exotic richness relationship across spatial scales: Favourable environments for native species are highly invasible. Functional Ecology. 2011;**25**:1106-1112. DOI: 10.1111/j.1365-2435.2011.01857.x

[48] Ramírez C, Finot V, San Martin C, Ellies A. El valor indicador ecológico de las malezas del centro-sur de Chile. Agrosur. 1991;**19**:94-116

Chapter 5

Game of Clones: Students Model the Dispersal and Fighting of Japanese Knotweed (*Fallopia japonica*)

Anneliese Fuchs, Christina Pichler-Koban, Wilfried Elmenreich and Michael Jungmeier

Abstract

Fallopia japonica as an invasive alien species in Europe and North America presents a significant problem to the existing flora as well as to infrastructures and agricultural land. That is why measures and attempts to control the plant are increasing rapidly. However, conservationists are not yet able to agree on the most suitable method. In the research project 'Game of Clones', a team of scientists together with the help of high school students is spatially modeling the spreading behavior of knotweed under different circumstances and is creating and providing a board game as well as a computer simulation as an experimental platform. To develop sustainable assumptions to be able to model the responses of knotweed to each control measure, a vast understanding of the plant is necessary. The chapter covers the results of research activities and experiments within the project and gives a comprehensive review about Japanese knotweed.

Keywords: Japanese knotweed, invasive species, dispersal, modeling, computer simulation, board game

1. Introduction

The spread of non-native species and their impact on the environment are a much-noticed topic in science and nature conservation. Recently, also a broader public is becoming increasingly interested, especially as the annual economic loss caused by alien species is estimated to be up to 5% of the world economic output [1]. Moreover, invasive alien species are an important factor in the loss of biodiversity. In fact, an analysis of the IUCN Red List shows that it is one of the most common threats associated with extinct species. Invasive alien species can also lead to changes in the structure and composition of ecosystems that have a significant negative impact on ecosystem services and affect the economy and well-being of humans [2–4]. Although the number of documented invasive species is underestimated in many countries, the introduction of invasive species has increased significantly. In Europe, for example, the number of invasive alien species increased by 76% between 1970 and 2007 (IUCN). Only a few of the thousands of species introduced into new areas actually become invasive, which is why their identification is the main objective of invasion biology. In Austria, 1110 alien vascular plant species have been identified, which account for 27% of the total Austrian flora. Of these, 17 species are problematic for nature conservation as they invade near-natural habitats [5]. Japanese knotweed, *Fallopia japonica*, is one of them and is considered to cause large changes to the communities and ecosystems it invades. Its large size and its clonal, monocultural growth lead to the visual, structural, and chemical transformation of ecosystems. Wherever the plant takes root, the diversity of plant species decreases. The remaining competing species are mostly non-native [6–9] and show strong reductions in height, biomass, and specific leaf area (SLA) [10]. Once a *F. japonica* stand is established, the clonal connectivity increases its ability to grow further [6]. The vast spreading in riparian areas also results in the reduction of an overall abundance of invertebrates [3, 7]. Therefore, a large-scale invasion of *Fallopia* species is likely to seriously affect the biodiversity and quality of ecosystems and should be prevented [7].

Not only does Japanese knotweed have a negative effect on the environment, but it also causes damage to infrastructure and costs effort and money for removal work. Each year, a considerable sum is spent on vegetation management on railway and road networks [11]. *Fallopia japonica* prefers manmade locations where other plants do not have a chance; in railway structures these are graveled areas, platforms, and loading areas. Weed control is primarily carried out in the track area in order to avoid fine soil and humus accumulation and thus reduce increased water retention capacity. Also, for the treatment of the track-accompanying paths, security is the main reason [12]. The urgent need for action can also be seen in our current projects: the project "Vegetation control on roads and railways" aims for vegetation control of traffic infrastructure areas with a balanced consideration between conventional and effective eco-alternative methods. In another project we are taking over the scientific monitoring for railway embankment grazing on the Koralm railway in order to control Japanese knotweed [13].

In agriculture, in addition to knotweed competing with crops, contaminated goods such as humus landfills pose a real problem. Open soils and disturbed vegetation provide an opportunity for problematic plants to colonize. One centimeter of root is enough for Japanese knotweed to form a new population [14]. According to Section 21 of the Carinthian Nature Conservation Act, the release or sowing of wild plants [...] into areas in which they are not native requires a permit. A permit may only be granted if neither the natural habitats nor the native wild animal and plant species are damaged. Large economic losses can therefore occur if humus landfills partly or fully overgrown with knotweed can no longer be used as such.

At present, there is no fully effective method to control knotweed. Still, in the literature, there is a long list of control methods ranging from mechanical methods such as pulling out and mowing [15] to grazing with sheep and goats [16], planting competitive native species [17, 18], covering the roots with tarpaulin, and using herbicides [18, 19] to biological control such as the use of Japanese knotweed psyllid [20].

In summary, characteristics, effects, and control measures of Japanese knotweed are subject to numerous research projects in Central Europe and North America; "Game of Clones" is one of them and approaches the topic in a somewhat different and playful way. A team of scientists together with the help of high school students aims to spatially model the spreading behavior of knotweed under different circumstances and to create and provide a computer simulation as an experimental platform as well as a board game. Considering that multiple components are required, first, a vast understanding of knotweed, especially regarding its ecological optima, its dispersal strategy, and its response to different control measures, is necessary. Therefrom, sustainable assumptions can be developed to be able to model the responses of knotweed to each control measure. For answering some of the questions, experiments will be used. The outcomes will lead to the creation of a board game and a computer simulation model based on a cellular automaton to be able to analyze and demonstrate the spreading behavior of knotweed in an interactive manner. Players will try out different measures to eradicate the clones and to keep particularly valuable areas clear from the weed. Doing this, they should go as easy on resources as possible. Depending on the individual starting points, different measures and combinations of measures will lead to success, in other words, reduce or stop the plant growth. The game takes place on actual existing land (satellite images), so the computer simulation can also be consulted for concrete action planning. The students in the research project will also play a part in the browser-based programming of the strategy game; in this way, they will simultaneously be an important reference group regarding its user-friendliness and functionality. The present chapter covers the results of the research activities and experiments within the project and gives a comprehensive review about Japanese knotweed. Section 2 starts with a description of all methods used, and Section 3 will be about the corresponding results. Section 4 finally discusses the question of the necessity of invasive species removal, summarizes the results, and concludes with a range of further recommendations for improving the existing evaluation and monitoring frameworks.

2. Methods

In "Game of Clones", a multitude of methods have been and are used. This is especially important because it is the only way to fully understand Japanese knotweed in all its parts and behaviors. The following chapter will describe each method with all its limitations and challenges in detail to be able to relate to the results.

2.1 Literature research

As a start, the team of researchers has carried out an extensive literature search. The contributions and articles collected were reviewed and classified as more or less relevant to the research question of the project. With the support of the Regional Museum of Carinthia (Landesmuseum Kärnten), a bibliography of over 200 relevant papers on *Fallopia japonica* was compiled and divided into various topics: classification, taxonomy, identification, characteristics, history, growth, reproduction, spreading, usage, impacts, monitoring, control, management, invasions, and modeling.

2.2 Phenotypic and genotypic identification

For a serious discussion about the plant, the most urgent question that needs to be clarified and cannot be answered by the literature is what exact species we are dealing with in our project area. In Central Europe, there is evidence for two introduced species, *Fallopia japonica* and *Fallopia sachalinensis*; their hybrid *Fallopia* × *bohemica* has begun spreading as well [6, 21, 22]. The two original species are relatively easy to distinguish based on the shape and size of their leaves, but discriminating them from hybrids is challenging, even for experts. Hence, we will make use of DNA-barcoding, a taxonomic method for species identification using the DNA sequence of a marker gene [23]. The sequence of base pairs is used as a marker for a particular species, analogous to the barcode on food packaging. Since the DNA sequence changes by point mutations at a generally uniform rate, more closely related individuals (and species) have more similar sequences. As long as a species remains undivided, i.e., has a common gene pool, differences between different populations are compensated again and again by gene flow. So, if samples from two individuals have clearly different sequences, this is a sign that they come from different species [24]. The analysis of two marker genes (chloroplast marker and nuclear marker) should provide information on hybridization and distribution of the species in our project area of the Austrian federal states of Carinthia and Styria. The chloroplast marker is inherited from the maternal organism, so by using it we will see what species was maternal. The nuclear marker will indicate if the plant is homo- or heterozygote, therefore a hybrid.

In the months of July and August 2018, 95 leaf samples were collected and sent to the Canadian Centre for DNA Barcoding (CCDB) in Guelph for DNA sequencing. For 3 weeks, 72 of the leaf samples were taken from individuals in Carinthia and Styria. Care was taken to ensure that different locations and morphologically different stands were chosen. If a site was selected, a tissue piece with an area of 1 × 0.5 cm was sampled with clean forceps. Preference was always given to the youngest and greenest parts of the plant, rich in plastids and meristematic cells such as the tip of a leaf. The samples were then placed in airtight bags of silica gel and kept to dry. Before proceeding to the next sample, it was crucial to ensure that no residual tissue remained on the forceps by rinsing them in 95% ethanol and wiping them with a clean absorbent paper. For each sample, a herbarium voucher of several leaves and flowers was collected, dried, and archived in the Regional Museum of Carinthia. Additional metadata included the assumed species, age, and sex as well as a detailed description of the site consisting of GPS coordinates, address, and site conditions. A photo documentation comprising location, entire plant, leaf surface and underside, and flower complements the sample collection (**Figures 1** and **2**).

Each sample was assigned to a Museum ID, which links it to the voucher, the metadata, and the photo documentation. A total of 13 of the 95 samples were collected from reliably identified individuals of all three species from the herbarium in the Regional Museum of Carinthia to serve as a reference. Ten of the 95 samples were not taken from the field, but CCDB offered to organize reference samples from Eastern Asia to have some samples from Japanese knotweed's native range.

The analysis is still ongoing; in case the sequencing will be successful, the data will be fed into the global Barcode of Life Data System (BOLD).

Figure 1.
Required tools for field sampling (E.C.O Institute of Ecology).

Figure 2.
Herbarium voucher of a sample of Fallopia sachalinensis (Herbarium collection code: KL—Kärntner Landesherbar).

2.3 Transect monitoring

Growth rates and propagation patterns are crucial parts of the basic data needed for the modeling of *Fallopia* populations. That is why we set up two transects on the campus of the Lakeside Science & Technology Park, a science and technology park in Klagenfurt. A transect is a straight line along which one counts and records occurrences of a species. The main advantage of transect mapping is its repeatability and standardization even under difficult terrain conditions. Both our transects (10 m each) were border on infiltration areas. The exact position of the transects was chosen in such a way that the shoots are rather in the middle of the observation area in order to be able to measure the propagation better. All methods were implemented according to the manual of vegetation-ecological monitoring by Andreas Traxler [25]. The transects were divided into 10 subplots (1 × 1 m each); the measured plants were each marked with a piece of yarn. In a weekly monitoring (April–July), the growth and propagation of *Fallopia japonica* was observed with two methods. On the one hand, three shoots were selected in both transects, in which the shoot width (at a height of 10 cm) and the height itself were measured with a caliper and meterstick. On the other hand, the number of shoots in each subplot was counted, and new shoots were marked and measured for their exact position. The data gain significance if they are interpreted in connection with the weather data for the period in question, as it is intended for the compilation of the logarithms of the computer simulation.

2.4 Rhizome uncovering

A good understanding of the underground processes in the *Fallopia* clone is of central importance for our research. To understand the connection between plant growth above- and underground and the knotweed's reaction to obstacles, we laid bare the entire root network of two stands in a large-scale field experiment. The method we used was already developed and successfully applied for the root exposure of forest trees in the past [26].

The stands are located on the campus of the Lakeside Science & Technology Park in Carinthia that borders directly on the Natura 2000 site Lendspitz-Maiernigg. During construction works 2 years ago, building rubble was piled up and populations of Japanese knotweed were able to colonize the area. The first location is a 4 m high hill with a 2-year-old stand; the second location borders on the parking lot, and its stands already exist for 4 years.

After the excavation work had been carried out and the site on the hill and next to the parking lot had been dug down by 2 m, the manual excavation work began. Together with the students, teachers, and soil experts of our two cooperating schools "BORG Spittal" based in Carinthia and "HBLFA Raumberg-Gumpenstein" based in Styria, the roots were then uncovered in a period of 2 days (**Figures 3** and **4**). The rough work was done with shovels, spades, and picks; the fine work was mainly done with screwdrivers. Bit by bit, the earth was dug away along the rhizomes and roots, thus exposing the roots. The results were documented in writing, in photographically, and in overview and detail drawings. The excavated shoot and rhizome parts were disposed of by the waste management department of the city of Klagenfurt so as not to contaminate further soil. After the two work days, the holes were dug up again by the excavator.

2.5 Rhizoboxes

Rhizoboxes are a non-invasive investigation method, which offers the possibility to survey the root system growth dynamics in time and space. Based on the root uncovering in Carinthia and the knowledge gained about length and width growth of the underground biomass, the experimental arrangements for the rhizoboxes were proposed. After a test experiment, adaptations took place; further experiments will follow. The method of using rhizoboxes aims to answer the following questions: how quickly do the rhizomes of knotweed grow (growth rates and depth and width growth) in vertical and horizontal rhizoboxes? What are the limiting factors (e.g., aboveground biomass, drought, cold, light, etc.)? For this purpose, 10 rhizoboxes in size of 30 × 100 cm were built, five in horizontal and five in vertical alignment (**Figures 5** and **6**).

Figure 3.
Excavation work at location 1 (E.C.O. Institute of Ecology).

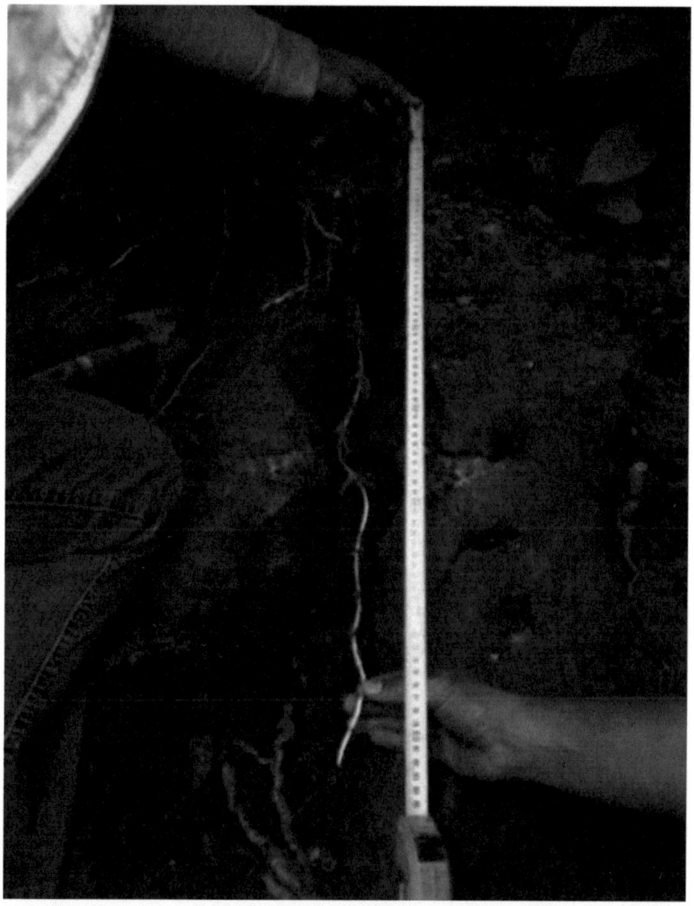

Figure 4.
Measuring the length growth of the rhizome (HBLFA Raumberg-Gumpenstein).

Figure 5.
The earth material must be sieved before the boxes are filled (HBLFA Raumberg-Gumpenstein).

Figure 6.
The rhizome was traced to simplify the measuring (HBLFA Raumberg-Gumpenstein).

Attempt 1: For the first experiment, fresh rhizome mass of Japanese knotweed was used to illustrate the growth in length, height, and width. On 13 July 2018, the first two boxes were filled with fresh earth material and the rhizome mass was planted (box 1 = 4 cm piece and box 2 = 7 cm piece). The first attempt was aborted because of glass jumps, mold formation, and too much soil and water.

Attempt 2: The second attempt started on July 27, 2018; four boxes were filled with fresh earth material. The length growth of the rhizome was measured approx. every 3–4 days, and the growth spurts were documented in an Excel file. The alignment of the boxes was optimized; the rhizome parts cast less intensively. After about 3 weeks, the growth directions and lengths were traced with a white marker. The boxes are still filled, and the rhizomes are moving inwards. Next steps will include:

- leaving some earth and rhizome material in the boxes and storing it over the winter (covered with fleece) in order to test whether there will be further growth next year,

- experimenting with two boxes being filled and sampled and simulating a longer growing period in the boarding school at a nearly constant temperature—the results are then evaluated in spring 2019, and

- starting a new rhizobox experiment with fresh material in spring 2019 and precisely documenting length and width growth.

The rhizobox experiments were all conducted by the students Philipp Poier and Julian Heywood and their teachers Renate Mayer and Irene Sölkner from the HBLFA Raumberg-Gumpenstein. Next year, observations will be longer and more regular. The present studies served as first pilot experiments to get familiar with the method.

2.6 Board game and computer simulation

All experiences gained in our research will influence and have influenced the development of both the analog and the digital version of the strategy game 'Game of Clones'. The game is based on a spatial model using cellular automatons (**Figure 7**) to display dynamic vegetation patterns [27]. The basic approach of cellular automatons is a subdivision of the area into equally sized, mostly quadratic fields. The dynamics of the modeling results from an interaction between the neighboring cells, in which a "state" is set to overlap the neighboring field (discretely modeled temporal development). When defining neighborhoods for quadratic cells, the van Neumann neighborhood and the Moore neighborhood are being distinguished. Whereas in the van Neumann neighborhood only cells with common edges are considered neighbors (this results in 4 neighbors per cell), the Moore neighborhood also defines diagonally adjacent cells as neighbors (this results in 8 neighbors per cell). For "Game of Clones", we chose the approach of a model with hexagonal cells, in which there are always exactly six neighbors. An application of a hexagonal model can be found, for example, in the SCIDDICA model, which models the behavior of landslides during strong water accumulations [28].

The modeling of the game includes biological parameters such as nutrient uptake, growth, and propagation rates as well as system parameters such as the shape and size of the cells to be simulated. This intersection of disciplines requires a close cooperation between expert biologists and modelers. Starting from the literature and empirical findings (reference area and experiments), the model is developed in an iterative process.

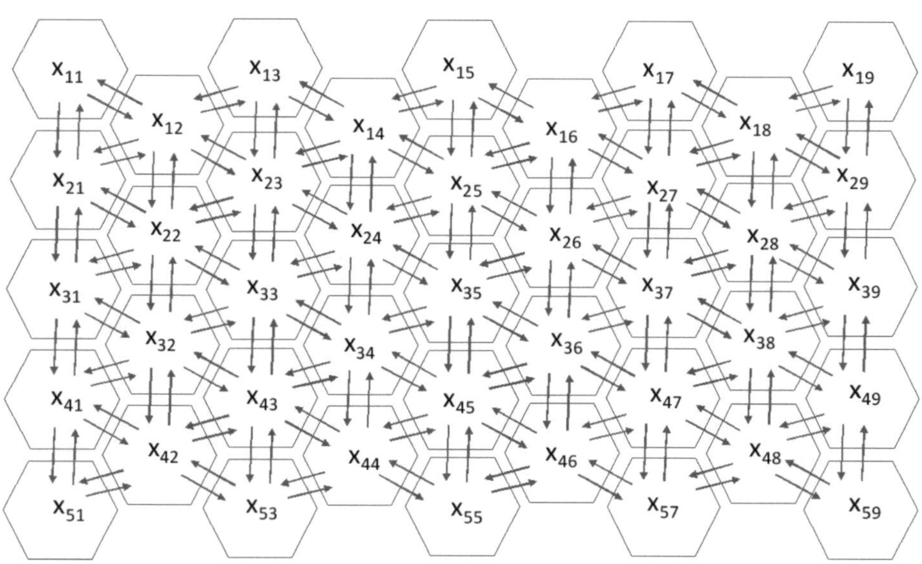

Figure 7.
Principle of the cellular automaton with hexagonal cells (Institute of Networked and Embedded Systems, University of Klagenfurt).

To be able to run the model with as many systems as possible, a browser-based implementation using html5 is provided to make the system compatible. Html5 supports the execution on operating systems and is—with certain restrictions with regard to screen size—also suitable for mobile devices. NetLogo, a multi-agent programming language with an integrated modeling environment, will be used for the simulation. The development process requires a repeated feedback of the results with biologists, whereby the model parameters and assumptions are repeatedly adjusted and compared with available findings (literature and experiments). This process is of particular scientific interest and value. The user interface is developed at the same time as the model is created. For this purpose, early user tests prototypes of the user interfaces to ensure ease of operation and an attractive design. The separation of model, view, and controller (Model-View-Controller Design Paradigm) allows a largely independent further development of program parts and supports a later independent use for other projects. The software is developed under an open source license and made available as a project result.

The board game "Game of Clones" is the analog version of the computer simulation and focuses on playability and fun instead of enforcing fully realistic scenarios. In the cooperative game, players work together in order to compete against Japanese knotweed, either winning or losing as a group. The board game was developed during biweekly meetings of the experts of the E.C.O. Institute of Ecology and the Institute of Networked and Embedded Systems from the University of Klagenfurt. A prototype of the board game is already available. During the development process, the students of BORG Spittal played through several test rounds and made a strong contribution to improving the game. The computer simulation will be completed by October 2019. In contrast to the board game, full attention will be paid to the closeness to reality whereby the program will be filled with all recorded data.

3. Results

3.1 Literature research

Fallopia japonica (Houtt.) from the knotweed family (Polygonaceae) has a number of synonyms, which makes literature research more difficult (frequently: *Polygonum cuspidatum* (Sieb. & Zucc.), *Reynoutria japonica* (Houtt.), and *Polygonum japonicum* (Meissn.)) as well as a number of phenotypically similar species and hybrids (in Austria in particular: *F. sachalinensis* and *F.* × *bohemica*) [29–31]. Until the definitive identification of the species in our study area, we use *Fallopia japonica* as the provisional collective name for these species.

Screening the literature, one of the main findings was that considering that it is only one single species, there is a huge amount of papers that revolve around Japanese knotweed. The articles cover various aspects of the plant, having a focus on morphology, systematics, spread, and control. Following a brief story-time about knotweed's introduction into Europe, this subchapter will be about the findings which proved to be relevant for the project.

Fallopia japonica was first introduced to Europe in 1825 by Philipp von Siebold, a Bavarian physician who worked for the Dutch government in Japan. Von Siebold had a strong interest in botany and natural history and sent a large shipment of live plants—over 500 different species—from Japan to the Netherlands, one of them being Japanese knotweed (under the name *Polygonum sieboldii*). It was intended to make a career as an ornamental and cattle feed plant and to be used in forestry as a feeding ground for red deer and as a covering plant for pheasants. The career as a useful plant did not start so well: it is of little use as a cover for pheasants, since it loses its leaves in winter and red deer do not eat it, neither our grazing livestock. However, since in early autumn it is an excellent bee pasture when most of the European native plants have already flowered, the beekeepers have discovered Japanese knotweed for themselves [32]. Although the German Federal Nature Conservation Act and most of the Austrian Federal State Conservation Acts prohibit the planting of alien plants in the wild, beekeepers generously distributed the Japanese knotweed in the area—a first step on the way to a spread that currently places Japanese knotweed at No. 37 in the "Global Invasive Species Database", a database managed by the Invasive Species Specialist Group (ISSG) of the IUCN Species Survival Commission.

When Japanese knotweed got introduced into Europe, they only introduced a female (male sterile) individuum, never the male indiviuum. A significant proportion of knotweed in Central Europe is not *F. japonica*, but the hybrid between it and *F. sachalinensis*—*F. bohemica*. This hybrid can reproduce with either parent and thus can replace the missing male specimens of *F. japonica*. In the same process, the hybrid produces the genetic diversity that *F. japonica* lacks so strikingly [33, 34].

All *Fallopia* species have a strong clonal growth, which allows them to surpass the surrounding species as well as to colonize new areas quickly. The basic unit of the rhizome system is a shoot clump that varies in size in different *Fallopia* species. In general, the apex of a rhizome branch eventually becomes an aerial shoot. When the shoot clump no longer produces new aerial shoots and dies, some lateral buds break the dormancy and begin to grow horizontally as new rhizome branches sometimes extending over 1 m. While *F. japonica* has rather large shoot clumps connected by long thin rhizomes, *F. sachalinensis* produces smaller shoot clumps that are more closely connected and grow in rows. *F.* × *bohemica* combines the characteristics of both parents and has an intermediate patch structure with smaller shoot clumps than *F. japonica* and longer rhizome connections between individual shoot clumps than *F. sachalinensis.* The fragmentation and spread of rhizomes by flooding or human activity are the most important means of propagation, as rhizome fragments of 1 cm length and 0.7 g weight can regenerate. *Fallopia* species can also regenerate from stem parts, but with lower regeneration rates. *F.* × *bohemica* had the highest regeneration rate of all taxa (61%) and is the most successful in regenerating and establishing new shoots. *F. japonica* and *F. sachalinensis* show lower regeneration rates (39 and 21%, respectively) [35].

The ability to regenerate in very poor soils with low nutrient requirements allows the plant to occur in a variety of habitats. It is not unusual for *F. japonica* to grow at the foot of buildings or on concrete surfaces [36]. The plant achieves it competitive superiority primarily by limiting access to light [37]. A factor, Japanese knotweed is very sensitive to, is frost. The plant is exposed to significant damages by late spring frosts when the shoots appear and by early frosts in autumn when the leaves senesce at the end of the growing season. This situation suggests that minimum spring temperatures may limit its range expansion [38]. However, climate change will open up habitats within threshold values, and frost conditions in these areas will be less severe and restrictive [39].

All these circumstances and many more make it extremely hard to get control over the invasive species. In the literature, there are many control methods and attempts described, but there are none that are completely convincing, and it amounts to a combination of different methods. Mechanical regulations focus on mowing, and although mowing during the vegetation period reduces the height and the diameter growth of shoots, the total weight of the biomass more or less stays the same [15]. The combination of cutting or mowing and using glyphosate has shown to be the most efficient and least time-consuming strategy so far [19, 40]. It is important to replant the area immediately with competitive native species to fight against invasive recolonization. On average, a suppression of knotweed is necessary for 2 years, before native species can be successfully established [18]. Another option is a long-term grazing of cows, sheep, or goats to keep the area knotweed-free [16]. A strategy which is becoming more popular is biological control, not least since Japanese knotweed was introduced without all its natural enemies. *Aphalara itadori* (*itadori* being the Japanese word for *Fallopia japonica*), a species of psyllid from Japan which feeds on Japanese knotweed, is the subject of an application for release into the wild in Great Britain. It has been licensed by the UK Government for the biological control of Japanese knotweed in England; this is the first time that biological control of a weed has been sanctioned in the European Union [41]. Other biological controls include a leaf beetle, *Gallerucida bifasciata* [42] or snails, *Succinea putris*, and *Urticicola umbrosus* [43].

3.2 Phenotypic and genotypic identification

So far, the leaf samples have been sent to the Canadian Centre for DNA Barcoding (CCDB), the analysis is not yet complete though. The expected results should clarify which *Fallopia* species occur in Carinthia and Styria and in what proportion. The phenotypic determination suggests that only sporadic samples of *F. sachalinensis* are expected. Due to insufficient morphological differences, the phenotypic discrimination between *F. japonica* and *F. x bohemica* was not possible.

3.3 Transect monitoring

The weekly monitoring of the two transects provided information on growth rates in height and diameter. The average height growth of the six plants studied decreased as the vegetation period progressed. While the growth of the shoots in April was averagely 21.7 cm per week, it dropped to an average of 14 cm in May and to an average of 1.2 cm in June. As a percentage, the plants grew by 40, 1, 15, and 0.9%, respectively. In **Figure 8**, the growth rate is visualized for each individual, and the initial length of the shoots is as follows: 117.6 cm (A.4.1.), 35 cm (A.5.1.), 77.5 cm (A.6.1.), 32 cm (B.6.1.), 26 cm (B.9.1.), and 36.7 cm (B.9.1.). The average diameter growth was between 1.2 and−0.5 mm per week. The negative values result

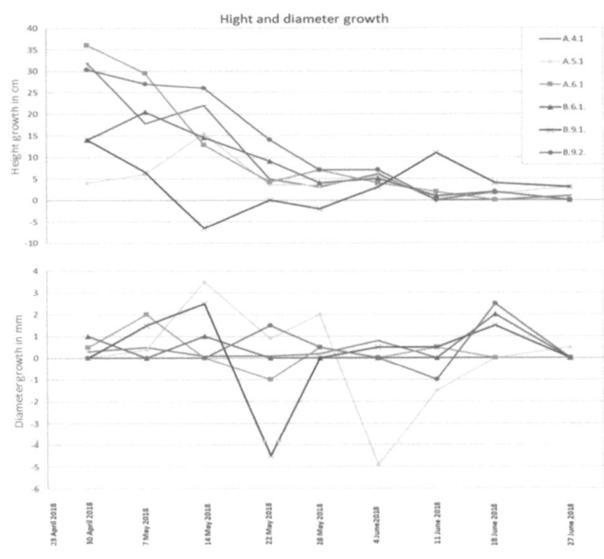

Figure 8.
Height and diameter growth of shoots in two transects (A&B).

from the fact that knotweed does not have woody shoots and the diameter size depends strongly on the water balance of the plant. Thus, it can happen that the diameter shrinks temporarily. While the growth of the diameter in April was averagely 0.3 mm (5%) per week, the growth rose to 1.8 mm (17%) in May. The plants lost biomass in the end of May/beginning of June resulting in negative values of −0.25 mm (−33%) and recovered in June with 1 mm (13%) growth rate. The initial diameters of the shoots as visualized in **Figure 8** are the following: 12 mm (A.4.1.), 2.1 mm (A.5.1.), 8.5 mm (A.6.1.), 3 mm (B.6.1.), 3 mm (B.9.1.), and 6 mm (B.9.1.). B.9.1. has low values starting in May; this results from a sudden wind break in the shoot, which has shortened the shoot and weakened the plant.

3.4 Rhizome uncovering

The rhizome uncovering could not confirm the assumption that the largest biomass of Japanese knotweed is underground. At site 1, the 2-year-old stand grew at a height of 4 m—the longest rhizomes reached a depth of 80 cm and were mainly horizontal. The reason to assume is that the plant mainly invests in the above-ground mass in the first few years. Site 2, a 4-year-old stand, underlines this assumption. The rhizomes reach 2 m into the deep until they stand at the groundwater body, which is generally high in Klagenfurt.

3.5 Rhizoboxes

The initial length of the rhizome pieces put in the rhizoboxes ranges from 5 to 24 cm. The results show that there is no correlation between initial rhizome length and growth rate. The boxes have been positioned horizontally and vertically, which showed a slight advantage for the rhizomes put in the horizontal boxes. All rhizomes started growing at a slow pace and speeded up at the end. These preliminary results have been conducted by students; further experiments are planned (**Figures 9–11**) and (**Table 1**).

3.6 Board game and computer simulation

All data and experiences gained in the project result in the development of a computer simulation. Along the way, we also created a board game for children and adults from the age of 10 years, which is currently on its way to a game publisher. The cooperation game allows players to work together as teammates against the opponent, Japanese knotweed. The game starts with a landscape full of differently

Figure 9.
Overview drawing Site 1 (E.C.O. Institute of Ecology).

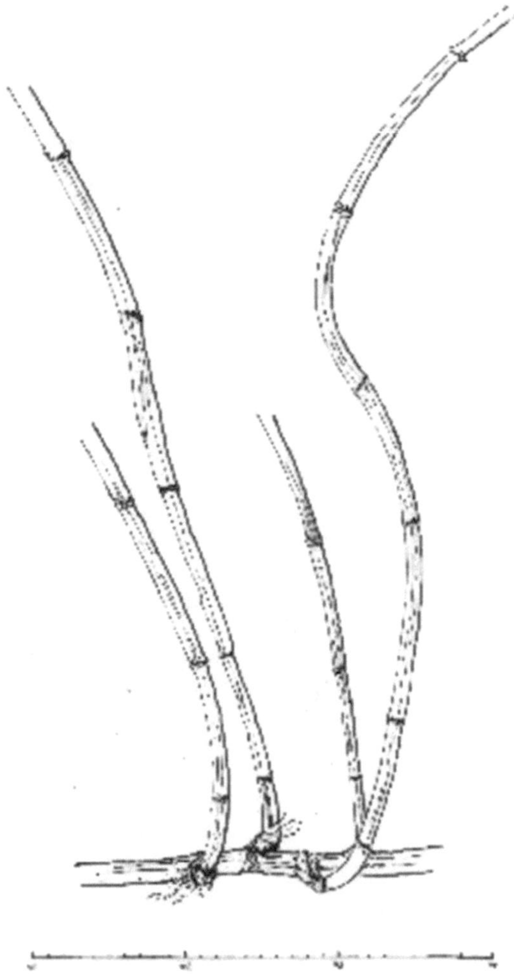

Figure 10.
Detail drawing of a rhizome (E.C.O. Institute of Ecology).

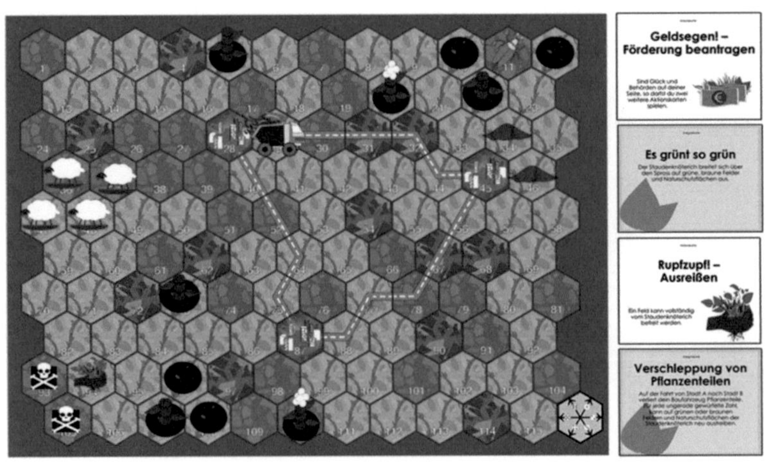

Figure 11.
Board game version of "Game of Clones".

Date	Rhizobox 1	Rhizobox 2	Rhizobox 3	Rhizobox 4
27 July 2018	5	5	24	4
30 July 2018	6	7	27	5.5
2 August 2018	6	7.5	30.5	5.5
6 August 2018	6	7.5	31	5.5
10 August 2018	7	8.5	31.5	6.5
31 August 2018	13.5	15.5	42.5	9.5

Table 1.
Rhizome length growth in rhizoboxes (rhizome pieces in cm); rhizoboxes 1 and 2 are positioned horizontally, and rhizoboxes 3 and 4 are positioned vertically (HBLFA Raumberg-Gumpenstein).

suitable habitats for knotweed, occupied by randomly distributed *Fallopia* clones. The players will try out different measures to eradicate the clones and to keep particularly valuable areas clear from the weed. Doing this, they should go as easy on resources as possible. The player team wins if they manage to displace all plants from the game plan and loses if one of the nature conservation areas is overgrown or destroyed by clones of knotweed. The game is based on event and action cards. Each round starts with an event card, meaning Japanese knotweed moves in a specific speed and a specific spreading mechanism. Then, it is the players' turn and they can choose between action cards that portray control methods such as mowing, pulling out, sheep grazing, glyphosate, cover foil, or biological control (*Aphalara itadori*). Depending on the individual starting points, different measures and combinations of measures will lead to success, i.e., reduce or stop the plant growth.

In several test rounds, we could see that the players started to realize how fast and determined Japanese knotweed can spread and how little can be done about it, if one does not take it seriously. The only way is to cooperate, to combine control measures and to act as soon as possible. Whenever the population is little, it is still quite easy to get rid of, and once the board is mostly overgrown by knotweed, it is extremely hard to push back the plant. The game is designed close to reality, and in terms of controlling knotweed, it shows that mechanical methods are time-consuming and inefficient, and that poison and cover foils are more efficient, but that they are not resource-saving and that one has to live with the consequences. Thus, Game of Clones creates awareness of invasive species in a playful way. The digital version is still in process, and the students in the research project will also play a part in the browser-based programming; in this way, they will simultaneously be an important reference group regarding its user-friendliness and functionality.

(E.C.O. Institute of Ecology, Institute of Networked and Embedded Systems at the University of Klagenfurt)

To our knowledge, the presented game is the only board game applying a cellular automata model to depict the spread of invasive plant species. While modeling plant growth with cellular automata is a well-established approach (a good overview can be found in [27]), there are only a few examples for usage of cellular automata simulations in board games: Franzel describes the usage of a board game to assess farmers' preferences among alternative agricultural technologies in [44]. Kang et al. [45] depict a computer simulation model addressing evolutionary game theory within a five-species jungle game, which is based on a Chinese board game. The work most related to our approach is the board game "Alien Invaders!" which teaches students how introduced species can affect native species with the example of native birds being affected by introduced species [46].

4. Conclusion

The project at the time of this publication still has a duration of 1 year, which means that many of the results are not yet complete and require further research.

The current results show that the issue of Japanese knotweed is very complex, and numerous studies and research projects have already been carried out and many are still ongoing. Due to the complexity and the costs of the control, the question arises whether the extensive control of knotweed is really necessary. But it turns out that even if one is of the opinion not to additionally intervene in the ecosystem and let nature take its course, *Fallopia japonica* also has significant economic effects, which cannot be ignored [1]. The need for further studies also arises from several disagreements in the literature such as the one regarding the gender of the species. Some say that the plant is clearly dioecious with distinct male and female individual organisms, and others speak of the plant being gynodioecious, which is the existence of male sterile and hermaphrodite individuals. There is not so much literature on the underground growth of Japanese knotweed. Hence the root uncovering was very informative, which is exactly why it would be advantageous to uncover the roots at another site, especially with older stands in order to present comparisons and observe the underground growth after the initial years.

The one method to fight knotweed does not exist. For every area, every situation, and every circumstance, a different strategy makes sense and mostly only the combination of different methods achieves an impact. When combating invasive species, however, one must always think in years. The computer simulation "Game of Clones" will be able to be underlaid with satellite images in order to establish a relationship to real areas. On the basis of the simulation, control strategies can be considered in advance of a measure concept. The project will also result in a practical guide and an explanatory video because the best way to combat an invasion is prevention and environmental education. During the collection of samples, there were numerous encounters with neighbors who were not aware of the problem and planted Japanese knotweed as a screen or threw plant remains into the compost. We hope that as many people as possible can be picked up by the game and sensitized to this topic. It would also be interesting to test whether the cooperative game method could be applied to other invasive plants in an adapted form.

Acknowledgements

This research project is part of Sparkling Science, a research program funded by the Austrian Federal Ministry of Education, Science and Research (BMWFW), a program that supports projects in which pupils **of all levels of education** are actively involved in the research process.

In the case of our project, students aged between 14 and 18 from BORG Spittal—a grammar school with special emphasis on music, art, sport, and science in Carinthia—supported us in the research project. Hence, special thanks to the students who helped with the sample collection for the DNA barcoding, with uncovering the root network, and with the transect monitoring and who also significantly contributed to improving the strategy game through test games and feedback rounds.

Special thanks also to our project partner DI Arthur Pitman from the Institute of Networked and Embedded Systems at the University of Klagenfurt. He was mainly responsible for co-developing the rules of the strategy game, processing the data gained in the project, and programming and modeling the computer simulation.

We express our warm thanks to the students Philipp Poier and Julian Heywood and the teachers and experts Dr. Andreas Bohner, DI Renate Mayer, and Irene Sölkner from the HBLFA Raumberg-Gumpenstein, a higher technical education and research institute in Styria that participated as a project partner. The school supported the rhizome uncovering and fully overtook the work of the rhizoboxes. Additionally, both students conducted their pre-scientific work on Japanese knotweed; within this framework, they carried out further experiments.

A big thank you also to Lisa Schmied, E.C.O. Institute of Ecology, who, during the rhizome uncovering, drew the rhizome of Japanese knotweed in all its perspectives.

We would also like to show our gratitude to our final project partner, Dr. Roland Eberwein from the Regional Museum of Carinthia (Kärntner Landesmuseum) who has been instrumental in researching the literature and who is responsible for archiving the herbarium vouchers from the DNA Barcoding.

Finally, many thanks to the Lakeside Science & Technology Park for supporting us and letting us set up transects and carry out a large rhizome uncovering on their property.

Conflict of interest

The Authors declare that there is no conflict of interest.

Author details

Anneliese Fuchs[1*], Christina Pichler-Koban[1], Wilfried Elmenreich[2] and Michael Jungmeier[1]

1 E.C.O. Institute of Ecology, Klagenfurt, Austria

2 Institute of Networked and Embedded Systems (NES), University of Klagenfurt, Austria

*Address all correspondence to: fuchs@e-c-o.at

References

[1] Pimentel D, McNair S, Janecka J, Wightman J, Simmonds C, O'Connel C, et al. Economic and environmental threats of alien plant, animal and microbe invasions. Agriculture, Ecosystems and Environment. 2001;**84**:1-20

[2] van Kleunen M, Dawson W, Essl F, Pergl J. Global exchange and accumulation of non-native plants. Nature. 2015;**525**:100-107. DOI: 10.1038/nature14910

[3] Horáčková J, Juřičková L, Šizling AL, Jarošík V, Pyšek P. Invasiveness does not predict impact: Response of native land snail communities to plant invasions in riparian habitats. PLoS One. 2014;**9**(9):10

[4] Gioria M, Dieterich B, Osborne B. Battle of the giants: Primary and secondary invasions by large herbaceous species. Biology & Environment Proceedings of the Royal Irish Academy. 2011;**111B**(3):1-17. DOI: 10.2307/23188047

[5] Essl F, Rabitsch W. Neobiota in Österreich. Wien: Umweltbundesamt; 2002. p. 432

[6] Aguilera AG, Alpert P, Dukes JS, Harrington R. Impacts of the invasive plant *Fallopia japonica* (Houtt.) on plant communities and ecosystem processes. Biological Invasions. 2010;**12**:1243-1252. DOI: 10.1007/s10530-009-9543-z

[7] Gerber E, Krebs C, Murrel C, Moretti M, Rocklin R, Schaffner U. Exotic invasive knotweeds (*Fallopia* ssp.) negatively affect native plant and invertebrate assemblages in European riparian habitats. Biological Conservation. 2008;**141**:614-654. DOI: DOI 10.1016/j.biocon.2007.12.009

[8] Künzi Y, Prati D, Fischer M, Boch S. Reduction of native diversity by invasive plants depends on habitat conditions. American Journal of Plant Sciences. 2016;**6**:2718-2733

[9] Skubala P. Invasive giant knotweed (*Fallopia sachalinensis*) alters the composition of oribatid mite communities. Biological Letters. 2012;**49**(2):143-155. DOI: 10.2478/v10120-012-0016-1

[10] Mincheva T, Barni E, Siniscalco C. From plant traits to invasion success: Impacts of the alien *Fallopia japonica* (Houtt.) Ronse Decraene on two native grassland species. Plant Biosystems—An International Journal Dealing with all Aspects of Plant Biology. 2016;**10**:1348-1357. DOI: 10.1080/11263504.2015.1115437

[11] Williams F, Eschen R, Harris A, Djeddour D, Pratt C, Shaw RS, et al. The Economic cost of invasive non-native species on Great Britain. CABI Knowledge for Life. 2010;**CAB/001/09**:198

[12] Brandes D. Flora und vegetation der Bahnhöfe Mitteleuropas. Phytocoenologia. 1983;**11**(1):31-115

[13] OEBB. Nicht allein auf weiter Flur [Internet]. 2018. Available from: https://konzern.oebb.at/de/nachhaltigkeit/umwelt/nicht-allein-auf-weiter-flur [Accessed: 22-10-2018]

[14] Gelpke G. Problempflanzen [Internet]. ALN Amt für Landschaft und Natur Fachstelle Naturschutz. Zürich. 2012. Available from: http://www.urtenen-schoenbuehl.ch/dl.php/de/0ev23-wflz1s/Problempflanzen.pdf [Accessed: 23-10-2018]

[15] Končeková L, Šebová H, Pintér E. Evaluation of population regulation of invasive species *Fallopia x bohemica* by repeated mowing. Acta Horticulturae et Regiotecturae. 2014;**17**(1):13-15. DOI: 10.2478/ahr-2014-0004

[16] Landesanstalt für Umweltschutz (LfU). Kontrolle des Japan-Knöterichs an Fließgewässern. In: Erprobung ausgewählter Methoden. Stuttgart: Handbuch Wasser; 1994. p. 63

[17] Koce J. The effects of leaf extracts of crack willow (*Salix fragilis*) on the growth of Japanese knotweed (*Fallopia japonica*). Acta Biologica Slovenica. 2016;**59**(1):13-21

[18] Skinner RH, van der Grinten M, Gover AE. Planting native species to control site reinfestation by Japanese Knotweed (*Fallopia japonica*) In. Ecological Restoration. 2012;**30**(2):192-199. DOI: 10.3368/er.30.3.192

[19] Child L, Wade PM. The Japanese Knotweed Manual. Chichester: Packard Publishing Limited; 2000. p. 123 ISBN: 1085341-127

[20] Clements DR, Larsen T, Grenz J. Knotweed management strategies in North America with the advent of widespread hybrid bohemian knotweed regional differences, and the potential for biocontrol via the psyllid *Aphalara itadori* Shinji. Invasive Plant Science and Management. 2016;**9**:60-70. DOI: 10.1614/IPSM-D-15-00047.1

[21] Bzdega K, Agnieszka J, Książczyk T, Lewandowska A, Gancarek M, Sliwinska E, et al. A survey of genetic variation and genome evaluation within the invasive *Fallopia* complex. PLoS One. 2016;**11**(8). DOI: 10.1371/journal.pone.0161854

[22] Gaskin JF, Schwarzländer M, Grevstad FS, Haverhals MA, Bourchier RS, Miller TW. Extreme differences in population structure and genetic diversity for three invasive congeners: Knotweeds in western North America. Biological Invasions. 2014;**16**:2127-2136. DOI: 10.1007/s10530-014-0652-y

[23] Sun XQ, Bai MM, Yao H, Guo JL, Li MM, Hang YY. DNA barcoding of populations of *Fallopia multiflora*, an indigenous herb in China. Genetics and Molecular Research. 2013;**12**(3):4078-4089. DOI: 10.4238/2013.September.27.9

[24] Hebert PDN, Cywinska A, Shelley LB, deWaard JR. Biological identifications through DNA barcodes. Proceedings of the Royal Society B. 2003;**270**(1512):313-321. DOI: 10.1098/rspb.2002.2218

[25] Traxler A. Handbuch des Vegetationsökologischen Monitorings. Wien: Umweltbundesamt; 1998. p. 158

[26] Kutschera L, Lichtenegger E. Wurzelatlas Mitteleuropäischer Waldbäume und Sträucher. Graz: Stocker Verlag; 2002. p. 604

[27] Balzter H, Braun PW, Köhler W. Cellular automata models for vegetation dynamics. Ecological Modelling. 1998;**107, 9**:113-125

[28] Avolio MV, Di Gregorio S, Lupiano V, Mazzanti P. SCIDDICA-SS_3: A new version of cellular automata model for simulatin fast moving landslides. The Journal of Supercomputing. 2013;**65**(2):682-696. DOI: 10.1007/s11227-013-0948-1

[29] Alberternst B, Boehmer HJ. 2011:15. Available from: Online Database of the European Network on Invasive Alien Species – NOBANIS. www.nobanis.org

[30] Tiebre M, Hardy O, Mahy G, Vanderhoeven S. Patterns of hybridization and hybrid survival in the invasive alien *Fallopia* complex (Polygonaceae). Plant Ecology and Evolution. 2011;**144**(11):12-18. DOI: DOI 10.5091/plecevo.2011.444

[31] Tiebre M, Hardy O, Mahy G, Vanderhoeven S. Hybridization and morphogenetic variation in the invasive alien *Fallopia* (Polygonaceae) complex in Belgium. American Journal of Botany. 2007;**94**(11):1900-1910. DOI: 10.3732/ajb.94.11.1900

[32] Bailey JP, Conolly AP. Prize-winners to pariahs—A history of Japanese Knotweed s. l. (Polygonaceae) in the British Isles. Watsonia. 2003;**23**:93-110

[33] Bailey JP, Bímová K, Mandák B. Asexual spread versus sexual reproduction and evolution in Japanese Knotweed s.l. sets the stage for the "Battle of the Clones". Biological Invasions. 2009;**11**:1189-1203. DOI: 10.1007/s10530-008-9381-4

[34] Grimsby JL, Tsirelson D, Gammon MA, Kesseli R. Genetic diversity and clonal vs sexual reproduction in *Fallopia spp*. (Polygonaceae). American Journal of Botany. 2007;**94**(6):957-964

[35] Adachi N, Terashima I, Takahashi M. Central die-back of monoclonal stands of *Reynoutria japonica* in an early stage of primary succesion on Mount Fuji. Annals of Botany. 1996;**77**:477-486

[36] Bímová K, Mandák B, Pyšek P. Experimental study of vegetative regeneration in four invasive Reynoutria taxa (Polygonaceae). Plant Ecology. 2003;**166**:1-11

[37] Siemens TJ, Blossey B. An evaluation of mechanisms preventing growth and survival of two native species in invasive bohemian knotweed (*Fallopia × bohemica*, Polygonaceae). American Journal of Botany. 2007;**94**:776-783

[38] Baxendale VJ, Tessier TJ. Duration of freezing necessary to damage the leaves of *Fallopia japonica* (Houtt.) Ronse Decraene. Plant Species Biology. 2015;**30**:279-284. DOI: 10.1111/1442-1984.12068

[39] Beerling DJ. The Impact of Temperature on the Northern distribution limits of the introduced species *Fallopia japonica* and *Impatiens glandulifera* in North-West Europe. Journal of Biogeography. 1993;**20**:45-53

[40] Bímová K, Mandák B, Pyšek P. Experimental control of Reynoutria congeners: A comparative study of a hybrid and its parents. In: Brundu G, Brock J, Camarda I, Child L, Wade M, editors. Plant Invasion: Species Ecology and Ecosystem Management. 2001. pp. 283-290

[41] Djeddour D, Shaw R. The biological control of *Fallopia japonica* in Great Britain: Review and current status. Outlooks on Pest Management. 2010;**21**:15-18. DOI: 10.1564/21feb04

[42] Wang Y, Ding J, Zhang G. *Gallerucida bifasciata* (Coleoptera: Chrysomelidae), a potential biological control agent for Japanese knotweed (*Fallopia japonica*). Biocontrol Science and Technology. 2008;**18**:59-74. DOI: 10.1080/09583150701742453

[43] Podroužková Š, Janovský Z, Horáčková J, Juřičková L. Do snails eat exotic plant species invading river floodplains. Journal of Molluscan Studies. 2014;**81**(1):1-8. DOI: 10.1093/mollus/eyu073

[44] Franzel S. Use of an indigenous board game, 'Bao', for assessing farmers' preferences among alternative agriculture technologies. In: Tomorrow's Agriculture: Incentives, Institutions, Infrastructure and Innovations - Proceedings of the Twenty-fouth International Conference of Agricultural Economists. 2018. pp. 416-424

[45] Kang Y, Pan Q, Wang X, He M. A five-species jungle game. PLoS One. 2016;**11**(6). DOI: 10.1371/journal.pone.0157938

[46] Stracey C. Alien Invaders! A Board Game about the Threats Posed by Introduced Species. Science Scope. 2008;**31**(6)